MONASH University

Reproducing Spermatogenesis

Queenie V. Neri

M.Sc. in Clinical and Translational Investigations

Copyright notice

To my parents,

Nilchar and Victorina,

for your continued love and support.

Table of Contents

Chapter 1

Abstract

Germ cells isolated from testicular tissue or derived from embryonic stem cells represent a possible treatment option for children who have survived cancer and for azoospermic men. Self-renewal of spermatogonia stem cells is the foundation for sustaining spermatogenesis throughout life. In recent years, spermatogonia transplantation in mammals has been carried out successfully and, more recently, the appearance of germ cell lines derived from embryonic stem cells has rendered the treatment or prevention of azoospermia conceivable. In our laboratory, we have identified and labeled human spermatogenic cells at various steps of differentiation and carried out their xenogeneic transplantation into recipient gonads. In addition, by appropriately modifying culture conditions, we have been able to propagate mouse spermatogonia stem cells *in vitro*.

Declaration

This thesis contains no material which has been accepted for the award of any other degree or diploma at any university or equivalent institution and that, to the best of my knowledge and belief, this thesis contains no material previously published or written by another person, except where due reference is made in the text of the thesis.

Chapter 2

Background

The human ability to procreate has been decreasing consistently during the past half century amongst the general population (Stefankiewicz et al. 2006). Approximately 6% of men between the ages of 15 and 44 are infertile or that their fecundity is severely compromised (Wright et al. 2004; Wright et al. 2006). An estimated 15% of couples of reproductive age encounter a range of problems pertaining to fertility. Half (50%) of all infertility cases is directly attributed to the male portion. The absence of spermatozoa in the ejaculate is defined as azoospermia. An estimated 50% of all azoospermia cases are due to hypo-spermatogenesis. Complete germ cell aplasia accounts for about 25%, and the remainder of azoospermic cases are due to spermatogenic arrest (Chan and Schlegel 2000).

While hormonal disturbances and abnormalities in semen production represent the main cause of infertility, anatomic anomalies and chemical substances which appear in the environment may contribute only in few cases. Substances disturbing hormonal balance (endocrine disruptors) such as pesticides, dioxins, and organic solvents represent the highest danger (Stefankiewicz et al. 2006). Improper work conditions and exposure to harmful substances such as drugs, past infections of the reproductive systems together with inadequate diet, alcohol, smoking, and long-term stress exposure may contribute as well. Latest findings have highlighted the role of genetic factors such as Y-chromosome microdeletions that are highly represented in certain populations, and may influence the reproductive ability of men.

The development of intracytoplasmic sperm injection (ICSI) aiming at selecting a single viable spermatozoon to inseminate a mature oocyte, has been heralded as the best procedure to achieve a consistent fertilization and pregnancy outcome since its inception in 1992 (Palermo et al. 1992; Palermo et al. 1996). The ability of virtually any spermatozoon, independently from the morphology or functional characteristics to induce oocyte activation and consequent pronuclear formation has resulted in the application of ICSI for the treatment of azoospermic patients. In spite of these accomplishments achieved by assisted reproductive technology (ART) in treating male factor infertility, there are still limitations. In fact, fecundity cannot be accomplished in men afflicted by Sertoli-cell-only syndrome or spermatogenic arrest where testicular extraction have been unsuccessful in identifying residual spermatogenetic foci.

Current research regarding the isolation of germ stem cells to fuel *in vitro* proliferation and eventual maturation is an attempt to fill a growing need for an alternative source of male gametes (Tanaka and Baba 2005). Spermatogonia stem cells (SSCs) are the only cell type in the male gonads that retain the ability to self-renew and under certain stimulus, be driven into differentiation leading to meiotic and post-meiotic stages (Brinster 2002). SSCs are very unique as they are the progenitor cells leading to spermatozoa production while also displaying the ability to revert back to a more totipotent state comparable to embryonic stem cells (Dym et al. 2009). Mastering skills to culture and propagate spermatogonia stem cells *in vitro* may pave the way to an array of treatment aimed at treating several disorders in male infertility.

The results of the experiments reported in the present thesis will be compared with current literature to explore all facets of *in vitro* male germ cell culture and eventual spermiogenesis. We aim to convey an understanding of the therapeutic potential of spermatogonia stem cell culture technologies in combating male infertility as well.

Chapter 3

Introduction

3.1 Adults with cancer

Approximately 69,000 individuals in the US between the ages of 15-39 year were diagnosed with cancer in 2011 (SEER 18 2007-2011). Many of them would have required chemotherapy and/or radiotherapy regimens, antineoplastic, and cytotoxic agents that being highly gonadotoxic compromise the ability to procreate. Gonadal failure is one of the major consequences of cancer therapy, remarkably so after ionizing radiation. Several reports on cancer survivors highlights the increased emotional distress related to chances of infertility post-antineoplastic treatment (Schover et al. 1999).

Depending on the underlying disease, the age of the oncological patient, the type of therapeutic agent used to treat the cancer, the cumulative radiation doses used or the duration of the chemotherapy, between 10-100% of surviving cancer patients would have shown impaired semen parameters following treatment (Tournaye et al. 2004). It is estimated that between 15 and 30% of these cured cancer patients become permanently sterile (Schrader et al. 2001). Thus, the concern that many patients undergoing potentially sterilizing antineoplastic cancer treatments have not been appropriately counseled about their future chances of fertility. Even though ART can successfully restore the reproductive capacity of many of these cancer patients, oncologists and their associates are still not fully aware of its limitations (Zapzalka et al. 1999; Allen et al. 2003). In fact, this lack of awareness in the medical community seems to be the main culprit for causing untimely or missed opportunity of cryostorage of gametes in these patients.

In the ICSI era, any semen sample, even those comprising only of a few motile spermatozoa, may be considered for sperm cryopreservation, and this should be performed prior to any cytotoxic treatment. Since most patients will develop azoospermia 2 or 3 months following chemotherapy (Schrader et al. 2001), cryobanking semen during the first month should still be advocated to overcome this major collateral effect (Carson et al. 1991). However, patients should be informed that chemotherapy may induce genetic and epigenetic abnormalities in both the short- (Robbins et al. 1997) and long-term (Brandriff et al. 1994; Foresta et al. 2000).

Azoospermia can sometimes be encountered in patients during the diagnosis of testicular cancer — this as a direct result of factors related to the malignancy on spermatogenesis (Berthelsen and Skakkebaek 1983). At orchiectomy, these patients may be offered sperm harvesting by vasal, epididymal aspiration (Baniel and Sella 2001), or by testicular sperm extraction (TESE) (Schrader et al. 2002) in order to bank their gametes prior to chemotherapy. Overall, assisted reproductive technologies particularly with ICSI is recommended for couples whose male partners are diagnosed with cancer and cryobanked their specimen before, after, and even during antineoplastic treatment.

3.2 Prepubertal age cancer

About one in every 600 children will develop cancer before the age of 15 (Hawkins and Stevens 1996). However, as a result of better treatment options, today the cancer death rate in infants and children has decreased more than that for any other age group, with up to 75% of all cases now being curable. At present, one in 1000 adults in the second or third decade of life is a childhood cancer survivor (Hawkins and Stevens 1996), and it has been estimated that this proportion will rise to one in 250 (Bleyer 1990).

With childhood cancer being treated more successfully, the focus of treatment has shifted towards the quality of life after treatment, and one aspect commonly overlooked is the prevention of sterility. In adults and adolescents, semen banking or cryopreservation of testicular tissue before radiation/chemotherapy are valuable preventive measures that circumvent sterility after treatment. However, no such prevention is possible prior to puberty since no active spermatogenesis has occurred, and so an average of 30% of children cancer-free remain sterile in the long term (Schrader et al. 2001).

One possible remedy consists of spermatogonia stem cell transplantation, the procedure in which either testis germ cells or stem cell colonies can hypothetically be transferred from a fertile donor into the seminiferous tubules of an infertile recipient. This is a technique currently being tested in experimental animals prior to human application.

The first experimental attempt at autologous spermatogonia transplantation was conducted by Brinster and colleagues in 1994 when mouse testicular cells obtained after enzymatic digestion were directly injected into seminiferous tubules of a recipient mouse previously made sterile by busulphan injections (a strong alkylating agent) (Brinster and Avarbock 1994). A month after post-germ cell transplantation, nascent germ cell colonies were identified as chains of cells on the tubular basement membrane. These spermatogonia stem cells (SSCs) implanted in the tubules then differentiated into later spermatogenic stages. In a further experiment, higher colonization rates of germ cells were reported by enhancing the transplanted donor cell population (Brinster and Nagano 1998).

Following these demonstrations, xenogeneic germ cell transplantation was conducted with testicular cells from fertile transgenic rats injected into the seminiferous tubules of immunosuppressed mice (Clouthier et al. 1996). In this case, mouse Sertoli cells clearly were able to serve as supporting somatic cells in rat germ cell recolonization and eventual differentiation. Other species used for donor cell injection into mouse seminiferous tubules that achieved successful colonization and differentiation were hamster (Ogawa et al. 1999), rabbits, and dogs (Dobrinski et al. 1999).

An alternative to germ cell transplantation is tissue grafting. Successful tissue xenografting of cryopreserved testicular specimen has been reported between various species (Honaramooz et al. 2002; Schlatt et al. 2002; Shinohara et al. 2002) and non-human primates (Orwig and Schlatt 2005).

Testicular tissue grafting entails placing a piece of the tissue, approximately 1 mm^3 size, subcutaneously under the dorsal skin or the testicular bursa. However, the differentiation capacity of frozen spermatogonia showed xenografts of testicular tissues from immature rhesus monkeys as having a blockage in the spermatocyte stage (Jahnukainen et al. 2007). Similar results were observed in frozen-thawed immature human testicular tissue grafted on the peritoneal bursa of mice testicles (Wyns et al. 2008) where numerous pre-meiotic spermatocytes were observed, however, spermatid-like cells did not express meiotic or post-meiotic markers.

3.3 Severely compromised spermatogenesis

3.3.1 Spermatogenic arrest

Spermatogenic arrest in men occurs most frequently at the spermatogonium stage or during arrested meiosis, the latter being most common at the primary spermatocyte stage (Martin-du Pan and Campana 1993; Sasagawa et al. 2001). Meiotic arrest, at primary spermatocyte stage, is the most recurrent maturational block in men with non-obstructive azoospermia (Martin-du Pan and Campana 1993).

Successful *in vitro* differentiation of spermatogenic cells into spermatids would be an extremely attractive option to treat some forms of male infertility caused by spermatogenic arrest (Tesarik et al. 1999; Staub 2001; Parks et al. 2003). Various approaches to *in vitro* spermatogenesis have been attempted, these mainly focusing on testicular cell cultures that are directed at achieving male germ cell differentiation. Among them, tissue culture, organ culture, and co-culture systems have been investigated extensively to assess which system can obtain meiotic or post-meiotic differentiation of these male germ cells (Hue et al. 1998; Cremades et al. 1999; Staub 2001; Sousa et al. 2002; Tesarik et al. 2002; Tanaka et al. 2003; Tesarik 2004). Despite the report of a normal child being born after fertilization with germ cells from a man with maturation arrest at the primary spermatocyte stage (Tesarik et al. 1999), an effective procedure for the completion of human spermatogenesis *in vitro* still remains elusive.

3.3.2 Germ cell aplasia

Sertoli-cell-only (SCO) syndrome, also called germ cell aplasia, describes a condition of the testes in which germ cells are absent and only Sertoli cells populate the seminiferous tubules. SCO syndrome presents during the evaluation of azoospermia in couples having difficulty in initiating a pregnancy. These men typically present with infertility as the only sign most often recognized between the ages of 20-40 years, when they are found to have no spermatozoa in their ejaculate. While the male urological physical examinations are often unremarkable, the diagnosis is confirmed by testicular biopsy. The investigation on the basis for the general epididymal dysfunction is still ongoing and the etiology and pathogenesis of this process are currently undefined. Some examples of diseases and disorders characterized by SCO and Leydig cell hyperplasia are Klinefelter's Syndrome, mumps orchitis, deletion of the Y chromosome at the AZFa or AZFb regions, and men/children that had anti-neoplastic therapy with radiation or chemotherapy.

The prevalence of the SCO syndrome in the overall population is extremely low. Of the 10% of US couples affected by infertility, approximately 30% have a pure male factor as the underlying cause, and another 20% have a combined male and female factor. Although precise figures are difficult to obtain, less than 10% of these infertile men have SCO syndrome with no known effective treatment(s) at the present time.

Chapter 4

Materials & Methods

4.1 Study protocols

4.1.1 Mouse

The experimental protocol was approved by the Institutional Animal Care and Use Committee (IACUC) of Weill Cornell Medical College (IACUC 9804-498A).

4.1.2 Human

This research protocol was approved by the Institutional Review Board (IRB 0498-200) of the Committee for Human Rights at New York Presbyterian Hospital-Weill Cornell Medical College and all patients were required to sign an informed consent.

4.2 Germ cell isolation

4.2.1 Mouse

Adult testicular tissue obtained from B_6D_2-F_1 mice were initially placed in M2 medium until tunica albuginea removal. Testes were then placed in solution containing collagenase I (300U/ml) and incubated for 15 min at 37 °C with manual agitation. After incubation, testes were washed twice with Ca^{2+}, Mg^{2+} free PBS and then incubated in a PBS containing 250U/ml trypsin and 0.2% EDTA for 10min at 37°C. Trypsin activity was arrested by the addition of sera and testicular cells were gently released from testicular tissue through a narrow bored pipette. The dissociated testicular cell suspension was overlaid on 30% (v/v) Percoll® prepared in Dulbecco's PBS (D-PBS) containing 1% FBS (PBS/FBS) and centrifuged at 600 x *g* for 8 min. Sedimented cells (pellet) and cells in density gradient were collected and washed twice with enough amount of PBS/FBS to remove silica gel particles. Cells were then resuspended with PBS/FBS containing 10 mM HEPES, 1 mM sodium pyruvate, 1 mg/ml glucose and antibiotics (PBS-S) (Tanaka et al. 2004; Neri et al. 2006).

B_6D_2-F_1 female mice (7–11 wk) were introduced to a B_6D_2-F_1 stud of the same age. The group-housed females (containing 2-3 females and 1 male) should enter estrus within 72 hours of being exposed to the male. The female mice were checked for a copulation plug and if confirmed, moved into a clean cage and the approximate delivery date recorded considering that the gestation lasts between 19-21 days. Cages were checked twice a week and daily when nearing the 19th day of gestation. It was noted when the mouse delivered to be able to obtain the testes of 6-10 day old pups. Mice were housed in a 12h light:12h dark cycle at constant temperature and provided with food and water *ad libitum*.

Testes were then obtained from 6-10 day old B_6D_2-F_2 mice. After dissection of the tunica albuginea, the tubules were exposed to collagenase type I or IV at a concentration of 260U/ml for 15 min, followed by 0.25% trypsin/1mM EDTA digestion for 3-5min (Brinster and Zimmermann 1994; Kubota et al. 2004, 2004).

In adult mice of about 10-30 weeks old, spermatogonia stem cells embody only a small minority of cells in the testes. It has been approximated that only 0.03% of SSCs are present in all cell populations in testes of an adult mouse. On the contrary, only spermatogonia stem cells are present when neonatal mice are utilized for experimentation (Tanaka et al. 2006). In order to maximize the greatest yield of both germ cells and type A spermatogonia from testicular cells, it is preferential to use neonates of approximately 5-10 dpp due to a higher proportion of gonocytes that are present in younger mice (Nagano et al. 1999).

4.2.2 Human (Germ cell identification and transplantation experiments)

Testicular samples were placed in 1 ml of prewarmed HEPES-buffered Earles' medium (Gibco BRL, Grand Island, NY, USA) supplemented with 5% human serum albumin (HSA) (Irvine Scientific, Santa Ana, CA, USA), 1.6 mM $CaCl_2$ (Sigma Chemical Co, St Louis, MO, USA), 25 μg/ml DNase (Sigma Chemical Co), and 1000 IU/ml collagenase type IV (Sigma Chemical Co) (Crabbe et al. 1998). DNase was added to the incubation medium to prevent clotting of the resulting cell suspension due to the release of free DNA from apoptotic cells (Crabbe et al. 1997; Crabbe et al. 1998). Testicular tissue was exposed to collagenase and incubated at 37°C for 1 hour, taking care of mixing the suspension every 10 or 15 minutes to enhance enzymatic digestion. Large portions of tubular walls were removed with forceps. Digested suspension was centrifuged twice at 500 *g* for 5 minutes.

The supernatant was removed and the pellet was resuspended in 500 μl of sperm cell medium (Brinster and Avarbock 1994). This medium contains 132 mM NaCl, 7.8 mM Na_2HPO_4, 2.6 mM KCl, 1.1 mM KH_2PO_4, 0.1 mM EDTA, 5.5 mM glucose, bovine serum albumin (BSA) (5 mg/ml), DNase I (200 μg/ml), and trypan blue (0.4 mg/ml) at pH 7.4 (Gibco BRL, custom formula 0980173, Grand Island, NY, USA).

Trypan blue was used to monitor the injection of seminiferous tubules. Concentration of spermatogonia, spermatocytes, and spermatids was assessed in disposable counting chambers (Cell-Vu®, Millennium Sciences Corp., New York, NY, USA) at 200x with a phase-contrast microscope. The cell viability was ascertained by eosin Y test. Specific identification of the spermatogenic cells was carried out on Testsimplets® slides (Boehringer Mannheim, Lewes, East Sussex, GB) and observed at 600x or 1000x (Reis et al. 2000).

4.3 Spermatogonia stem cell enrichment

4.3.1 Magnetic-Activated Cell Sorting (MACS)

Magnetic microbeads conjugated to monoclonal rat anti-mouse Thy-1 (CD90) antibodies, which mainly bind to SSCs surface, were used for magnetic-activated cell sorting (MACS) to enrich Thy-1 positive cells from testicular cell suspensions. The single-cell suspension (0.38 - 2.7 x 10^6 cells in 90 µl of PBS-S) was incubated with 10µl of Thy-1 microbeads for 20 mins at 4°C. After incubation with Thy-1 antibody, cells were washed once with excess of PBS-S and resuspended in 1 ml PBS-S. Cell suspensions were loaded on the top of a separation column that was placed in a magnetic field. Thy-1 negative cells were collected as the first fraction passing through a magnetic field column. After removal of the tube from the magnetic field, the magnetically retained Thy-1 positive cells were eluted with PBS-S. For the identification of spermatogonia including SSC, cells obtained after enzymatic digestion and Percoll® centrifugation and Thy-1 positive (selected) and negative cells (unselected) after MACS were assessed for cell concentration in a counting chamber, or stained on Testsimplet® slides for evaluation of morphometric parameters (Tanaka et al. 2005).

4.3.2 EasySep

When using EasySep™, 100 µl of mouse cell suspensions in PBS/2% FBS/1mM EDTA were incubated at room temperature (RT) for 15 min with Thy-1 (CD90) antibody conjugated with phycoerythrin (PE) (Stem Cell Technologies, BC, Canada). Cells were incubated with EasySep™ PE selection cocktail at RT for 15 min, followed by 10 min incubation of cell suspension with EasySep™ magnetic nanoparticles at RT. After the addition of 1.0 mL buffer, the Thy1-magnetic particle suspension was transferred to a polystyrene tube, which was then placed in a magnet (EasySep™; as per the manufacturer's instructions). The purity of Thy-1+ cells was evaluated by the visualization of PE under the epifluorescent microscopy immediately after selection (Tanaka et al. 2005).

4.4 Spermatogonia stem cell *in vitro* culture

Thy-1 positively selected cells were plated in a chemically defined medium composed of 0.2% bovine serum albumin, iron-saturated transferrin, free fatty acids, H_2SeO_3, 2-mercaptoethanol, HEPES, Putrescine, glutamine, and antibiotics enriched with human glial cell line-derived neurotrophic factor (GDNF, 40 ng/ml) and human basic fibroblast growth factor (bFGF, 1 ng/ml) (Kubota et al. 2004). In some cultures, unselected testicular cells were plated on culture dishes that were supplemented with either 1) GDNF + bFGF, 2) human recombinant follicle stimulating hormone (FSH–which is known to stimulate GDNF in Sertoli cells, 1.4 IU/mL) (Tadokoro et al. 2002) + bFGF (Neri et al. 2006) or 3) GDNF + bFGF + Leukemia Inhibitory Factor (LIF).

4.5 Identification of germ cells

Mouse testes and cell aggregates in culture were fixed in 2% paraformaldehyde (PFA) and paraffinized. Sections of the paraffinized tissue (3 – 5 μm thickness) were stained with haematoxylin/eosin (H&E). Morphology and architecture of the germ cells in their seminiferous tubule context were also assessed by observation of H&E stained histological sections displaying their location in the lumen. The localization and morphology of germ cells and somatic cells in H&E stained histological sections were examined under a light microscope. Each cell was classified into PGC, SSC, spermatogenic or

somatic cells according to the morphological criteria, such as nuclear size and nucleus/cytoplasm ratio (Russell 1990).
Tissue sections were incubated with 20% goat serum in PBS for 60 min at room temperature, followed by incubation with the monoclonal rat anti-mouse Thy-1 antibody. After washing in a protein-supplemented buffer, specimens were incubated with FITC conjugated secondary IgG antibody for 60 min at room temperature. After further rinsing, nuclei were counterstained with dimidinopropylindol (DAPI) antifade solution and then assessed under a fluorescent microscope. Identification of spermatogonia was carried out by structural localization and cytological characteristics that allowed distinction from more advanced maturational stages and somatic cells.

Isolated and cultured cells were assessed for alkaline phosphatase (AP) activity, Thy-1 presence, VASA and Oct-4 expression to identify PGC, SSC, and the spermatogenic cells, respectively. Cells were fixed with 2% PFA blocked/permeabilized with PBS/BSA/goat serum/Triton X-10, then incubated with rabbit polyclonal VASA (Toyooka et al. 2000) or Oct-4 diluted with PBS/BSA/goat serum overnight at 4°C. After three washes with PBS/BSA, specimens were incubated with goat anti-rabbit Alexa Fluor 488 diluted in

1:200 with PBS/0.1%BSA for 1 hr at RT. Alkaline phosphatase activity in fixed cells with 2% PFA was detected using an azo-dye technique with a Texas-Red filter under fluorescence. The stain solution contained naphthol AS-MX phosphate (Sigma) and fast red TR salt (Sigma). The nuclei were counterstained with DAPI. Immunopositive cells were evaluated under epifluorescent microscopy.

After digestion, 20 μl cell testicular cell suspension was transferred to a 1.5 ml centrifuge tube. Two hundred μl of 4% PFA was added and incubated for 30 min at 4°C for fixation. After centrifugation at 600 *g* for 5 min the pellet was resuspended in 100 μl of PBS with 20% normal goat serum, which was incubated for 60 min at room temperature. After rinsing, the pellet was resuspended in 50 μl of monoclonal rat anti-human integrin-α-6 antibody (1:20 dilution), and was incubated for 60 min at room temperature. Cell suspension was rinsed by centrifugation and the pellet was resuspended in a buffered solution containing the secondary antibody. After removing the excess secondary antibody by centrifugation, an aliquot of 10 μl resuspended pellet was placed on a poly-L-lysine coated slide and covered with DAPI antifade solution.

4.6 Xenogeneic transplantation

4.6.1 Recipient Mice

Two different strains of mouse were used as recipient, the W/W^v and the SCID mouse. W/W^v male mice (Jackson Laboratory, Maine, USA is characterized by mutations that affect the structure of the c-kit receptor thus resulting in germ cell aplasia (Mintz 1957) while severely immunodeficient mice (SCID) lack B- and T-cells (Taconic, Germantown, New York). Endogenous spermatogenesis in the SCID mice was suppressed by intraperitoneal injection of the antineoplastic alkylating agent busulfan (Sigma Chemical Co) (32 mg/kg) at least 4 weeks prior to donor cell transplantation (Russell and Brinster 1996). W/W^v and SCID male mice were housed in a 12h light:12h dark cycle at constant temperature and provided with food and water *ad libitum*. The SCID mice were kept under specific pathogen-free conditions with all the food, water, and bedding autoclaved before use.

4.6.2 Transplantation Procedure

Human spermatogonia were directly injected into individual seminiferous tubules as previously described (Brinster and Zimmermann 1994). Mice were anesthetized by intraperitoneal administration of pentobarbital 40-70 mg/kg. Incision sites were shaved and disinfected with povidone-iodine solution. Testes were exteriorized through a midline abdominal incision, positioned to present superficial tubules.

Pipettes were made from borosilicate glass (Drummond Scientific, Broomall, PA, USA) drawn on a horizontal micropuller (Model: 753, Campden Instruments Ltd., Loughborough, UK) and calibrated, cut, and melted to a closed tip with a microforge (Model: MF-9/900, Narishige, New York/New Jersey Scientific Inc., Middlebush, NJ, USA) that were then chipped to produce a 40 μm sharp tip. The pipette was mounted on a pipette holder connected to a disposable 10 ml syringe with a teflon tube.

Under a dissecting microscope, the tunica albuginea was opened to expose the seminiferous tubules. An individual tubule was penetrated and germ cells released. The flow of the suspension and consequent filling of the tubule was monitored by the blue marker (Trypan blue). When increased resistance to the injection was perceived, another tubule was selected and the procedure repeated.

4.6.3 Isolation of Transplanted Testes

The recipient mice were maintained between 47 and 144 days following injection of human germ cells prior to sacrifice. This time interval would presumably allow transplanted cells to undergo 1 or 2 cycles of spermatogenesis required for a stem cell to evolve into a fully developed spermatozoa (Heller and Clermont 1964). Sacrifice of the host was performed by CO_2 overdose and the testes were surgically removed, weighed, and enumerated.

4.6.4 Assay for Implantation

Testes and epididymis were fixed in Bouin's fixative, sliced into slabs, dehydrated in a graded series of ethanol, infiltrated in xylene, and embedded in paraffin. Cross sections of 3 μm thickness were cut every 50 μm in each specimen and stained with H&E. Twelve to twenty histological sections of each testis including at least 100 circular tubules were examined for the presence of foci of recolonization within the germinal epithelium. To distinguish xenogeneic spermatogenesis from residual spermatogenesis, an immunocytochemical marker specific for human spermatogenic cells (4D4 anti-proacrosin monoclonal antibody; kindly provided by Dr. Denise Escalier) (Escalier et al. 1991) was utilized.

Immunolabeling was performed by the three-step immunoperoxidase technique using the biotin-avidin system (Dako Corporation, Carpinteria, CA, USA). Some sections were first treated with either 7.5% H_2O_2 in PBS for 10 minutes at room temperature or with 0.1% trypsin in 0.05 M Tris buffer (pH 7.6) with 0.1% $CaCl_2$ for a time not exceeding 15 minutes at 37°C. The reaction was stopped with 4°C PBS. To block non-specific antibody binding, all washed sections were incubated at room temperature for 20 minutes in either 5% normal horse serum or 5% BSA (grade V, Sigma Chemical Co) in PBS. Sections were then incubated with mAb 4D4 culture supernatant (diluted 1:100) in a humid chamber at room temperature for 1 hour. Sections treated with the biotinylated antibody were then incubated for 30 minutes with preformed avidin-peroxidase complexes, and amino-ethyl-carbazole (AEC) was used as chromogen. The sections were then counterstained with Harris' hematoxylin and mounted in an aqueous medium (Glycerged, Dakopatts, Santa Barbara, USA).

4.7 Data analysis

A χ^2 test was utilized for the different comparisons involving efficiency of germ cell isolation. A two-tailed test was used to assess significance, considered at 5% probability. Statistical comparisons were reported in text and tables only when significance was reached. All statistical computations were conducted using StatView 512+ (BrainPower Inc., Calabasas, CA, USA).

4.8 Equipment list

Description	Manufacturer	Model
Borosilicate glass pipettes	Drummond Scientific, Broomall, PA, USA	
Centrifuge	Eppendorf AG, Hamburg, Germany	5702
Microcentrifuge	Beckman Coulter Inc., Brea, CA, USA	Microfuge® 18
Cell-Vu® Sperm counting Chambers	Millenium Sciences Corp., New York, NY	DRM-600
Makler® Counting Chambers	Sefi-Medical Instruments Ltd, Haifa, Israel	
Fuchs-Rosen Counting Chambers	ProSource Scientific Inc., Ottawa, Ontario, Canada	
EasySep™	Stem Cell Technologies	
Epifluorescent microscope	Olympus Corporation, Shinjuku, Tokyo, Japan	BX60
Horizontal micropuller	Campden instruments Ltd., Loughborough, UK	753
Light microscope	Nikon Instruments Inc. Melville, NY, USA	TE200 DAPI+OT

Description	Manufacturer	Model
Microforge	Narishige, New York/New Jersey Scientific Inc., Middlebush, NJ, USA	MF-9/900
Syringe	Becton, Dickinson & Co., Franklin Lakes, NJ, USA	10 ml
Testsimplet® Slides	Boehringer Mannheim, Lewes, East Sussex, GB	
Stereomicroscope	Olympus	SZX12

Micromanipulator Workstation

Description	Manufacturer	Model
Inverted microscope	Olympus	IX-70
Motor coarse controller	Narishige	MM-188
Hydraulic micromanipulators	Narishige	MO-188
Microinjectors including tool holders	Narishige	IM-6
PiezoActuator	Burleigh	PiezoDrill
PiezoActuator	Primetech	PMM-150FU

Description	Manufacturer	Model
Spindle View System	CRI	SPV-001
ElectroCell Manipulator	BTX	ECM2001
Imaging software and filter set	Leica/Applied Imaging	CytoVision Genetic

Microtool set-up

Description	Manufacturer	Model
Micropuller	Campden	753*i*
Microgrinder	Narishige	EG-40
Microforge	Narishige	MF-900

4.9 Reagent list

Description	Manufacturer	Supplement (Concentration)
4-(2-hydroxyethyl)-1-piperazineethanesulfonic acid (HEPES) - buffered Earles' medium	Gibco BRL, Grand Island, NY, USA	

Description	Manufacturer	Supplement (Concentration)
4-(2-hydroxyethyl)-1-piperazineethanesulfonic acid (HEPES)	Sigma Chemical Co., St Louis, MO, USA	10 mM
4D4 anti-proacrosin mAb	Gift from Dr. Denise Escalier	
Antibiotics (Penicillin-Streptomycin solution)	Mediatech, Inc., Manassas, VA, USA	
Bouin's Fixative	Polysciences, Inc.,Warrington, PA, USA	
Bovine Serum Albumin (BSA)	Sigma Chemical Co., St Louis, MO, USA	5 mg/ml; 0.2%
Bovine Serum Albumin (BSA) grade V	Sigma Chemical Co., St Louis, MO, USA	1%
Busulfan	Sigma Chemical Co., St Louis, MO, USA	32 mg/kg
Calcium chloride ($CaCl_2$)	Sigma Chemical Co., St Louis, MO, USA	1.6 mM
Collagenase type	Life Technologies Corp., Carlsbad, CA, USA	260 U/ml

Description	Manufacturer	Supplement (Concentration)
Collagenase type 4	Sigma Chemical Co., St Louis, MO, USA	260 U/ml; 1000 IU/ml
DAPI	EMD Millipore Corporation, Billerica, MA, USA	
Disodium hydrogen phosphate (Na_2HPO_4)	Sigma Chemical Co., St Louis, MO, USA	7.8 nM
DMEM	Gibco	
DNase I	Sigma Chemical Co., St Louis, MO, USA	25 μg/ml; 200 μg/ml
Eosin	Sigma Chemical Co., St Louis, MO, USA	
Ethanol	Sigma Chemical Co., St Louis, MO, USA	100%; 80%; 60%; 40%
Ethylenediaminetetraacetic acid (EDTA)	Sigma Chemical Co., St Louis, MO, USA	0.1 mM; 1 mM
Fast red TR salt	Sigma Chemical Co., St Louis, MO, USA	
Fetal Bovine Serum (FBS)	Sigma Chemical Co., St Louis, MO, USA	2%

Description	Manufacturer	Supplement (Concentration)
Free fatty acids	Sigma Chemical Co., St Louis, MO, USA	7.6 μeq/L
Glucose	Sigma Chemical Co., St Louis, MO, USA	5.5 mM
Glutamine	Sigma Chemical Co., St Louis, MO, USA	2 mM
Goat anti-rabbit Alexa Fluor 488	Life Technologie Corp., Carlsbad, CA, USA	1:200
Goat serum	Sigma Chemical Co., St Louis, MO, USA	5%
Haematoxylin	Sigma Chemical Co., St Louis, MO, USA	
HTF-HEPES	Irvine Scientific, Santa Ana, CA, USA	
Human basic fibroblast growth factor (bFGF)	Sigma Chemical Co., St Louis, MO, USA	1 ng/ml
Human Glial cell line-derived neutrophic factor (GDNF)	R&D Systems Inc., Minneapolis, MN, USA	40 ng/ml
Human recombinant follicle stimulating factor (FSH)	R&D Systems Inc., Minneapolis, MN, USA	1.4 IU/ml

Description	**Manufacturer**	**Supplement** (Concentration)
Human Serum Albumin (HSA)	Irvine Scientific, Santa Ana, CA, USA	5%
Human Tubal Fluid (HTF)	Irvine Scientific, Santa Ana, CA, USA	
Hyaluronidase	Sigma Chemical Co., St Louis, MO, USA	100 U/ml
Iron-saturated Transferrin	Sigma Chemical Co., St Louis, MO, USA	10 μg/ml
$KSOM^{AA}$	Special media	—
Liquid nitrogen	TW Smith	
M2	Sigma	
Mercaptoethanol (2-mercaptoethanol)	Gibco BRL, Grand Island, NY, USA	10 mM
Monopotassium phosphate (KH_2PO_4)	Sigma Chemical Co., St Louis, MO, USA	1.1 mM
Naphtol AS-MX Phosphate	Sigma Chemical Co., St Louis, MO, USA	
Normal Horse Serum	Abcam PLC, Cambridge, MA, USA	5%
Oct-4	Abcam PLC, Cambridge, MA, USA	

Description	Manufacturer	Supplement (Concentration)
Paraffin	Sigma Chemical Co., St Louis, MO, USA	
Paraformaldehyde (PFA)	Sigma Chemical Co., St Louis, MO, USA	2%; 4%
PBS	Sigma Chemical Co., St Louis, MO, USA	
Pentobarbital	Sigma Chemical Co., St Louis, MO, USA	40-70 mg/kg
Polyclonal VASA	Rockland Immunochemicals Inc. Gilbertsville, PA	
Potassium chloride (KCl)	Sigma Chemical Co., St Louis, MO, USA	2.6 mM
Povidone-Iodine swabs	Cardinal Health, Dublin, OH, USA	
Putrescine	Sigma Chemical Co., St Louis, MO, USA	60 mM
PVP solution	Irvine Scientific, Santa Ana, CA, USA	7%
RPMI-125	Gibco	
Selenous acid (H2SeO3)	Sigma Chemical Co., St Louis, MO, USA	3x10-8M

Description	**Manufacturer**	**Supplement** **(Concentration)**
Sodium chloride (NaCl)	Sigma Chemical Co., St Louis, MO, USA	132 mM
Thy-1 (CD90) Ab coniugated with Phycoerythrin (PE)	Stemcell Techonologies, Vancouver, British Columbia, Canada	
Triton X-100	Sigma Chemical Co., St Louis, MO, USA	
Trypan Blue	Gibco BRL, custom formula 0980173, Grand Island, NY, USA	0.4 mg/ml at pH 7.4
Trypsin	Gibco BRL, custom formula 0980173, Grand Island, NY, USA	0.25%
Waymouth-IBMX	Gibco	
Xylene	Sigma Chemical Co., St Louis, MO, USA	

Chapter 5

Results and Discussion

5.1 Early Development and Germ Cell Determination

An early fundamental event during development is the divergence of germ cells from somatic cells. Germ cell specification in mammals takes place at the onset of

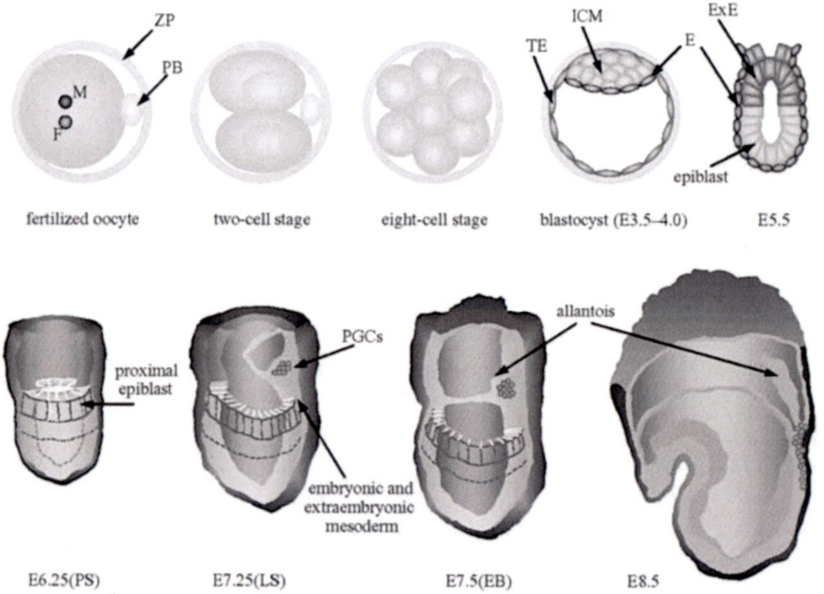

Figure 1. Schematic representation of early mouse development and appearance of PGCs. Mouse developments starts with the fertilization of an oocyte and following fertilization, the contents of the male and female pronuclei combine to from a reorganized diploid genome. Zygotic transcription is initiated at the 2-cell stage. The trophectoderm begins to form at the 8- to 16-cell stage, while inside a population of pluripotent cells, ICM forms by E3.5. At E4.5, the blastocyst implants and during E5.0-5.5, ICM cells cavitate to form the epiblast which is proximally contacted with the extraembryonic ectoderm. At E7.25, after gastrulation, the germ cells first become identifiable at the base of the allantois. At E8.5, most germ cells start migrating towards the genital ridges, where they undergo sexual differentiation and eventually produce either male or female gametes - adapted from (Downs and Davies 1993).

gastrulation, after implantation of the embryo.

In the mouse, primordial germ cells (PGCs) are first distinguished at the base of the allantois in gastrulating embryos at embryonic day (E) 7.25 (Ginsburg et al. 1990). Lineage studies of epiblast cells show that mouse PGCs are specified by inductive interactions at the onset of gastrulation (Figure 1) (Lawson and Hage 1994; Saitou et al. 2002).

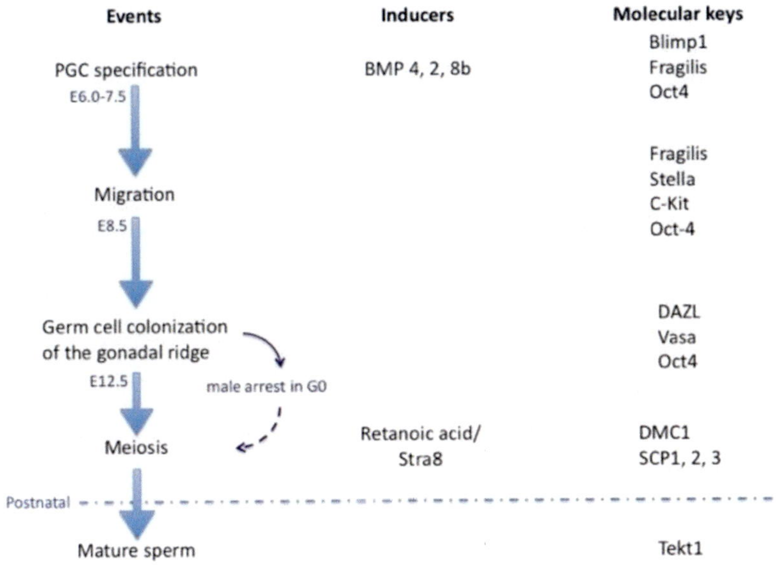

Figure 2. The germline development: events and molecular keys. During in vivo development, BMP family stimulates a subset of epiplast cells that begin to express Fragilis and Blimp1 and become PGCs that migrate towards the gonadal ridge.

Moreover, primary cultures of epiblast fragments from embryos at E5.5-E6.0 generate migrating PGCs when they are cocultured with extraembryonic ectoderm (Yoshimizu et al. 2001), and culturing of whole epiblasts from E6.0 embryos on feeder cells expressing both BMP4 and BMP8b gives rise to PGCs (Ying et al. 2001). These results reveal that BMPs derived from the extraembryonic ectoderm play crucial roles in PGC determination in the proximal epiblast. PGCs will then migrate in the developing embryo towards the gonadal ridge, proliferate, and differentiate into gonocytes between E8.5-12.5 (Figure 2).

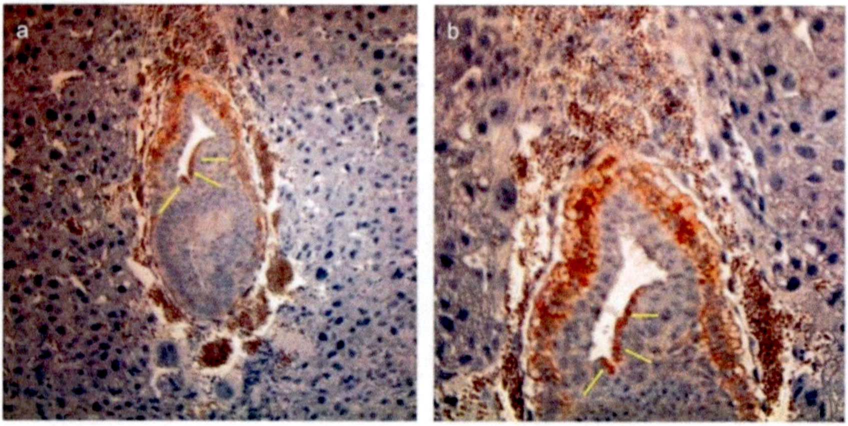

Figure 3. Perfused mouse embryo at 6.5 dpc. PGCs are labeled with VASA (yellow arrows) at (a) 100x and (b) 200x.

To study the morphology, appearance, and migratory behavior of PGCs, in preliminary experiments we assessed embryo development as early as 6.5 dpc

(Figure 3a,b) with the embryo being 120 x 250 μm in size with an implantation site of 2 x 3 mm.

After immunohistochemistry on 4-7 sections of 5 μm, a total of 14.6% (12/82) of the cells per embryo section expressed *fragilis* – a marker for premigratory PGCs. In a 12.5 dpc embryo (Figure 4) with a distinguishable gonadal ridge measuring 25.0 x 87.5 μm after H&E staining, sections of this region evidenced 25.9% (45/174) alkaline phosphatase activity and VASA positive cells (Figure 4) indicating post-migratory PGCs.

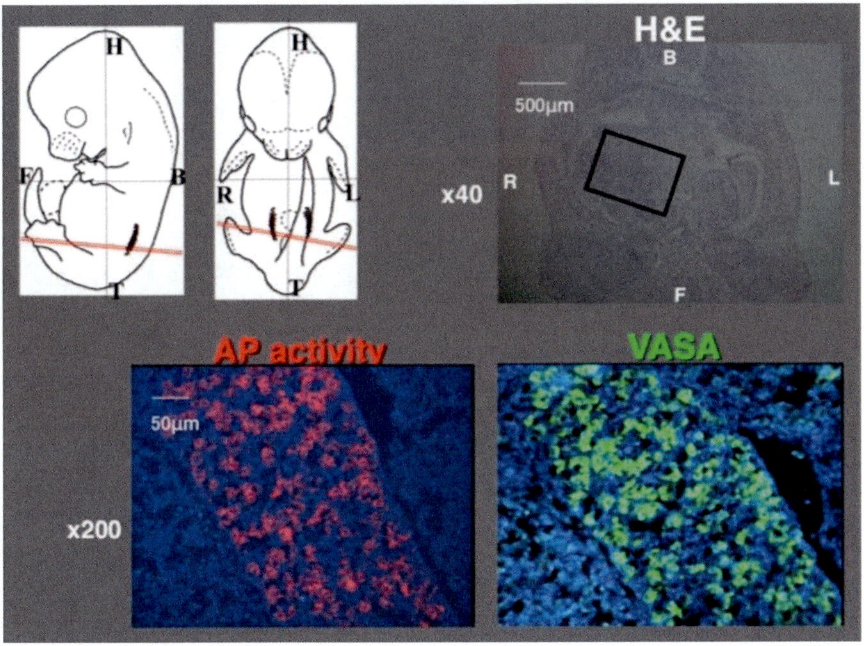

Figure 4. Perfused mouse embryo at 12.5 dpc. One cross-section was stained with H&E and another section was double labeled with AP activity (red) and VASA (green).

5.2 Seminiferous Tubule Mapping of Adult and Neonatal Mice

The increasing use of testicular specimens obtained from patients with non-obstructive azoospermia (NOA) involves an often frustrating search for a viable spermatozoon, or in the absence of spermatozoa, the problem of distinguishing immature germ cells from white blood cells, Sertoli cells, or interstitial cells.

In tissue sections, spermatogenic cells are easily identified by their morphology as well as architectural localization in the seminiferous tubules. However, once they are singled out from the tubules by either mechanical and/or enzymatic digestions, it is difficult to distinguish each maturational stage of each spermatogenic cell. Here we attempt to map the spermatogenic line in the seminiferous tubules of adult and neonatal mice.

Testicular samples were obtained from 2 month old B_6D_2-F_1 mice and 6 day old pups. Transverse sections of seminiferous tubules stained with H&E allowed architectural identification of germ cells and somatic cells because the entire spermatogenetic sequence was identifiable (Figure 5a,c). Localization of integrin-α-6 at the plasma membranes of the outermost layer of the germ cells confirmed the spermatogonia stage (Figure 5b,d). This allowed distinction from subsequent maturational stages where this marker was missing.

In another set of experiments, minced testicular tissue allowed

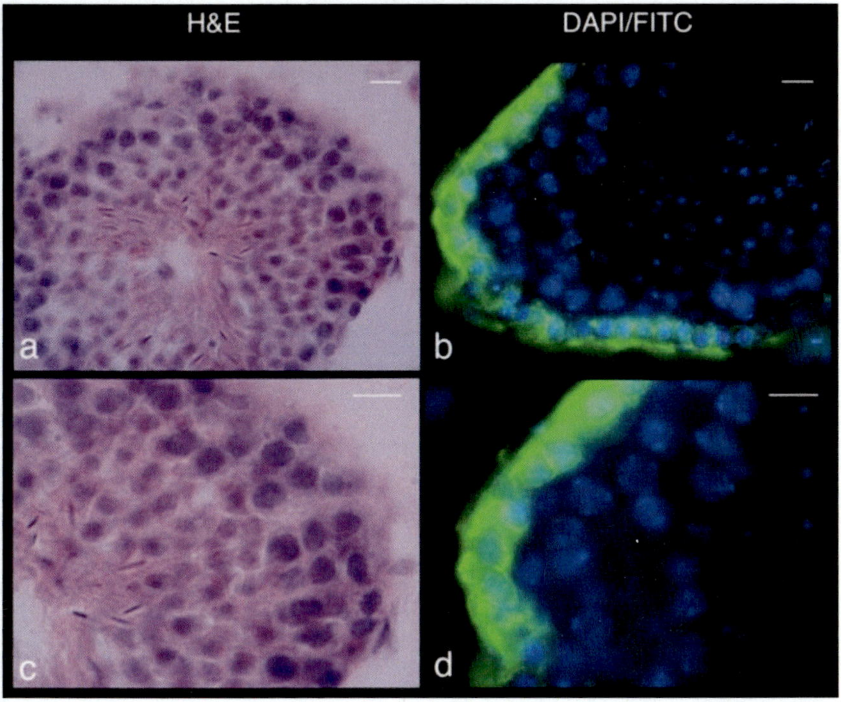

Figure 5. Histological section of mouse seminiferous tubules. H&E staining (a,c) represents complete spermatogenesis (600x and 1000x, respectively). Expression of integrin-α-6 (green signal) (b,d) on the plasma membranes of the outer layer of the germ cells (spermatogonia). Nuclei are counterstained with DAPI in blue (b,d). The more advanced maturational stages, from primary spermatocytes to spermatozoa, did not integrin-α-6 signals. Scale bars = 10 μm.

assessment of concentration, morphology as well as cellular integrity. The average germ cell concentration was estimated to be 66.9 x 10^6/ml while spermatozoa was 11.7 x 10^6/ml.

Germ cell integrity on a prestained slide after aggressive mechanical dissection was about 94% of the total cells. All maturational stages starting from spermatogonia to a mature spermatozoon could be identified as well as Sertoli cells (Figure 6). The proportions of the spermatogenetic stages (90.4%) as well as somatic cell type (9.6%) are depicted in **Table 1**.

Table 1. Proportion of spermatogenic cells at different maturational stages versus somatic cells

Testicular cells	**%**
Spermatogenic	
Spermatozoa	12.6
Spermatids	35.1
Spermatocytes	25.2
Spermatogonia	17.5
Somatic	
Sertoli	5.7
Leydig	2.1
Myoid	1.8
Total	100

Few Leydig and myoid cells were observed in the cell suspension. In addition, in the mince tissue integrin-α-6 assay allowed undoubtful identification of spermatogonia and Sertoli cells (Figure 6).

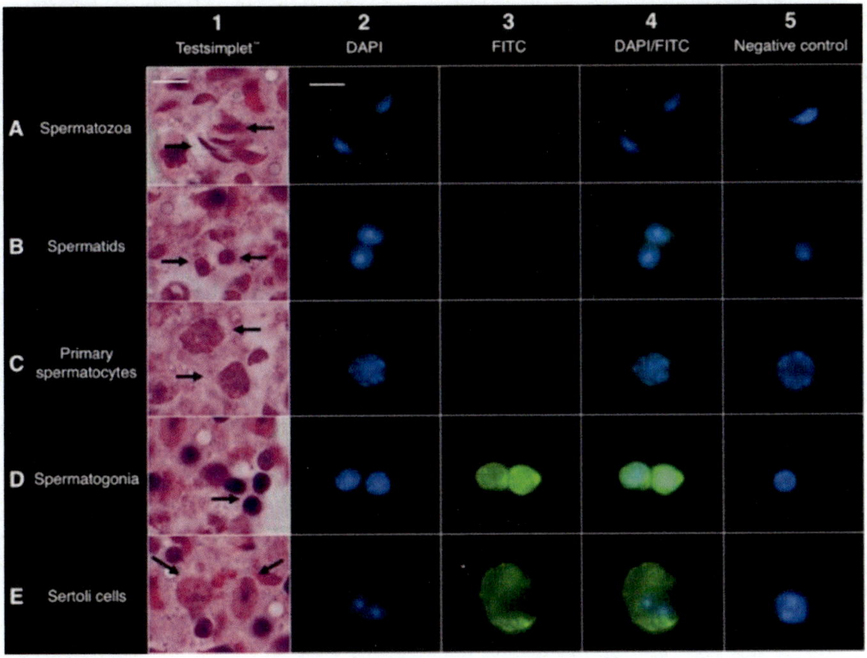

Figure 6. Staining of different spermatogenetic stages (A1 to D1) as well as Sertoli cells (E1). Column 2 as well as Column 5 evidence DNA staining. Column 3 identified the different germ cells and somatic cells stained by integrin-α-6 (green) without and with the visualization of the concurrent DNA staining. Only the spermatogonia stage and, weakly, the Sertolic cells display the green fluorescence (Column 4). Scale bar = 10 μm.

In adult mice of about 10-30 weeks old, spermatogonia stem cells embody only a small minority of cells in the testes. It has been approximated that only 0.03% of SSCs are present in all cell populations in testes of an adult mouse (see Section 5.3).

When adult and neonatal testes are compared, it is evident that the neonatal mice present only three cell types, spermatogonia, pre-Sertoli cells, and interstitial cells (Figure 7).

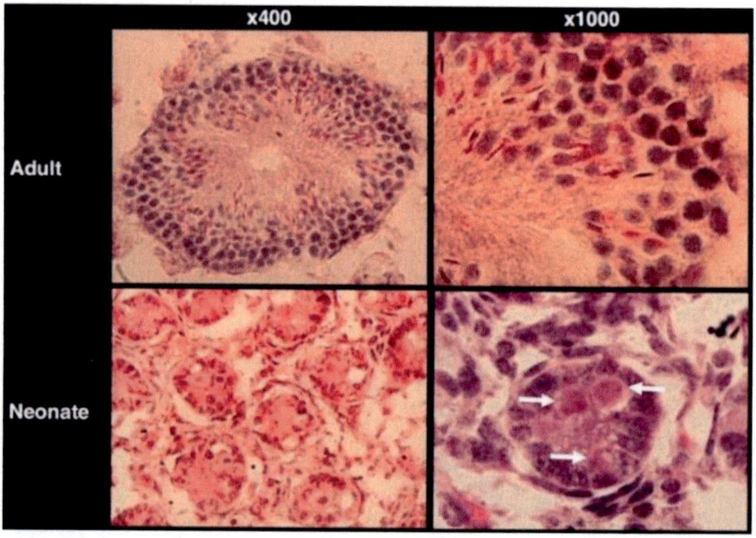

Figure 7. Histological section of adult and neonatal mouse seminiferous tubules. H&E staining evidenced an entire adult tubule (400x) with all the spermatogenic line (1000x). In the neonate, the same field at 400x evidenced multiple tubules and at 1000x, we identified three germ cells (white arrows) at the basement membrane of a single tubule.

5.3 Isolation, Identification, and Selection of Spermatogonia Stem Cells

In this section, we explored simple means of isolation and identification of SSCs in adult and neonatal mouse testicular tissues. We attempted to isolate and select spermatogonia stem cells by utilizing magnetic activated cell sorting (Figure 8).

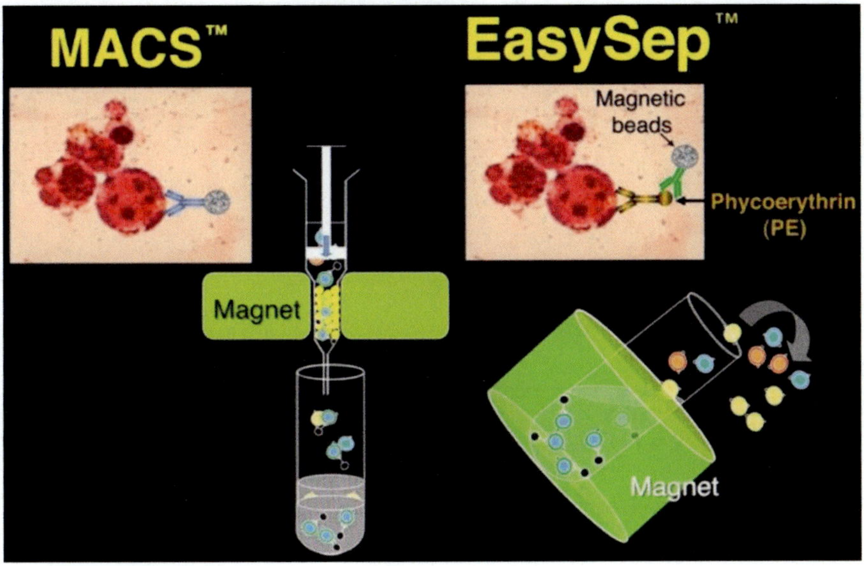

Figure 8. Cell suspensions were first processed by density gradient and then sorted either with MACS™ or EasySep™ after incubation with monoclonal Thy-1 magnetic-bead conjugated antibody.

In order to obtain the maximal yield of SSCs, testes were retrieved from both neonate and adult B_6D_2-F_1 mice (Figure 9).

	No. of				
Age	Mice	Testes		Size (mm)	Weight (mg)
Neonate (5-10 dpp)	21	42		1.6 x 1.3	7.6
Adult (12-28 wk)	6	12	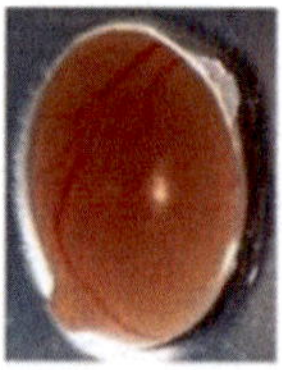	8.2 x 5.5	110.4

Figure 9. Specimen characteristics of neonate and adult testes. As expected, neonatal mice were smaller in size and weight.

To have a better understanding of the diverse cell architectural composition of neonatal and adult testes, we compared the number of round cells obtained after enzymatic digestion, and after sorting by MACS™ (**Table 2**).

Table 2. Recovery of testicular cells

No. of cells (%)	Neonate	Adult
Enzymatic digestion	0.58×10^6	102×10^6
Density gradient	—	1.8×10^6 (1.8)
Sorting	47,000 (8.1)	73,000 (0.07)
Expected (gonocytes/SSCs)	**15,000 (2.6)**	**34,000 (0.03)**

In the adult, the testicular cell suspension were layered onto a 30% density gradient and centrifuged. We were able to remove spermatozoa and other cellular debris, obtaining 1.8 million round cells. CD90 (Thy-1/Thy-1.2) antibody was elected as a marker to identify SSC fractions. The excess antibody was removed and the resulting cell suspension was loaded onto a separation column within a magnetic field. After cell sorting, the total germ cells obtained were 47,000 for neonates and 73,000 for the adult. The early maturational stage proportion such as gonocytes and SSCs was higher in the neonatal germinal epithelium.

The next step was to determine the SSC selection efficiency according to the two different magnetic cell sorting methods.

Table 3. Spermatogonia stem cell selection in adult testes

No. of cells (%)	MACS™	EasySep™	*P* value
Unselected	33 x10^6	29 x10^6	—
Selected	37,000 (0.1)	20,000 (0.07)	—
SSCs	20,0054 (54.2)	13,820 (69.1)	< 0.001

In both adult (**Table 3**) and neonatal testes (**Table 4**), the EasySep™ was able to recover more SSCs in comparison to the

MACS™. This difference in yield was more evident in the neonatal specimens since in the latter, there are only three cell populations, namely spermatogonia, pre-Sertoli cells, and interstitial cells. In fact, sorting of six dissected neonatal testes (795,000 testicular cells) followed by EasySep™ yielded about 43,500 CD90 positive cells with a > 80% enrichment rate ($P < 0.0001$).

Table 4. Spermatogonia stem cell selection in neonatal testes

No. of cells (%)	MACS™	EasySep™	*P* value
Unselected	0.77×10^6	0.53×10^6	
Selected	64,000 (8.3)	35,000 (6.6)	
SSC	298/505 (58.0)	418/502 (83.3)	< 0.001
SSC morphometry	293/502 (51.3)	415/510 (81.4)	< 0.001

The morphology of SSCs are characterized by a relatively large cell size of ~18μm and high nucleus/cytoplasm ratio. Morphology and architecture of the germ cells in their seminiferous tubule context were assessed by observation of H&E stained histological sections displaying their location in the lumen (Figure 7). After selection, some of the cell suspension was browsed under the fluorescent microscope to confirm the presence of SSCs (Figure 10).

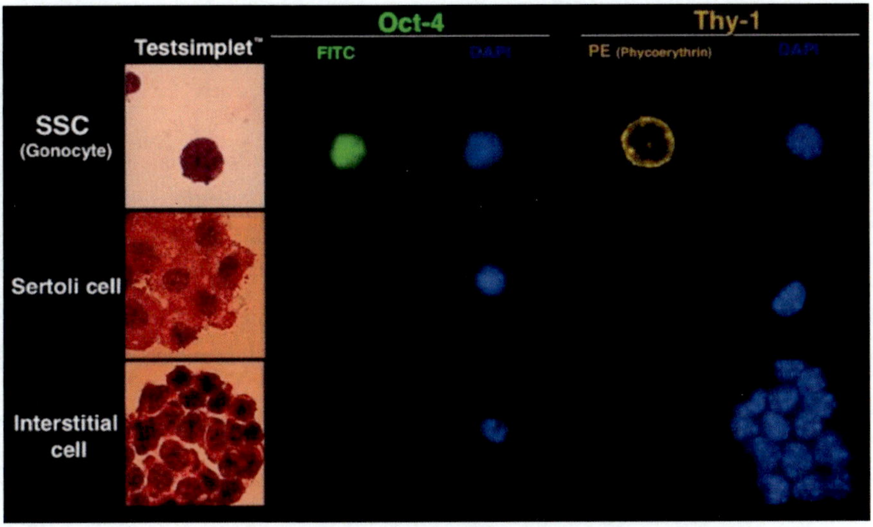

Figure 10. Neonatal testes after enzymatic digestion were placed on a Testsimplet® for morphological evaluation, indicating the presence of gonocytes, Sertoli cells, and interstitial cells. Germ cell identification was performed either by morphology or specific markers. Gonocytes (spermatogonia stem cells; SSC) are distinguished by their morphology of being a large cell with high nucleus/cytoplasmic ration as well as positive expression of specific markers such as Oct-4 and Thy-1, while the somatic cells are negative (DAPI only, in blue).

5.4 Media Optimization

Gonocytes/SSCs self-renew and produce daughter cells that commit to differentiate into spermatozoa throughout adult life. SSCs can be identified unequivocally by a functional assay using a transplantation technique in which donor testis cells are injected into the seminiferous tubules of infertile recipient males (Brinster and Avarbock 1994; Brinster and Zimmermann 1994). Under these conditions, only SSCs are able to generate foci of germ cells undergoing spermatogenesis (Brinster 2002).

One approach to the problem is cultivating SSCs under conditions that allow self-renewal and possibly stimulate differentiation. For this purpose, it is essential to establish a culture system with defined, experimentally modifiable characteristics. Glial cell line-derived neurotrophic factor (GDNF) is responsible for stimulating SSC renewal in rodents (Schmidt et al. 2009). In addition, several BMP genes are expressed in the testis, in particular, BMP7 and BMP8 a/b transcripts (Zhao et al. 1996; Zhao et al. 2001), suggesting that not only Sertoli cells, but also germ cells secrete molecules that can mediate paracrine interactions in the testis. *In vitro* co-culture experiments have shown that both BMP4 and BMP8B proteins function synergistically to induce PGC specification from the epiblast (Ying et al. 2001). BMP2, another member of the BMP family, acts in cooperation with BMP4 as a third signal in PGC formation (Ying and Zhao 2001).

In preliminary experiments, Enriched Thy-1$^+$ germ cells were seeded on gelatinized coated 60mm tissue culture dishes in standard medium supplemented with either fetal bovine serum (FBS) or serum-free (BSA only) along with additional components such as several growth factors (**Table 5**).

Table 5. SSC *in vitro* culture medium

Serum-free		**Serum**
	α-MEM	
0.2% BSA		15% FBS
	Insulin (5 μg/ml)	
	Transferrin (10 μg/ml)	
	Free Fatty Acids stock (7.6 μeg/L)	
	Sodium Selenate (3 x 10^{-8} M)	
	2-mercaptoethanol (50 μM)	
	HEPES (10 μM)	
	Putrescine (60 μM)	
	L-glutamine (2 μM)	
	Antibiotics (50 mg/ml)	
	LIF (1000 U/ml)	
	GDNF (40 ng/ml)	
	βFGF (1 ng/ml)	

We observed that there were 2.3X more SSC colonies in the medium containing 15% serum (Figure 11).

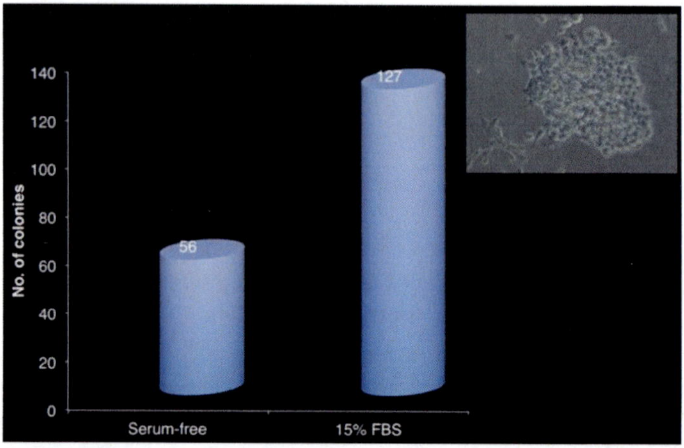

Figure 11. Number of SSC colonies (image inset) according to serum supplementation.

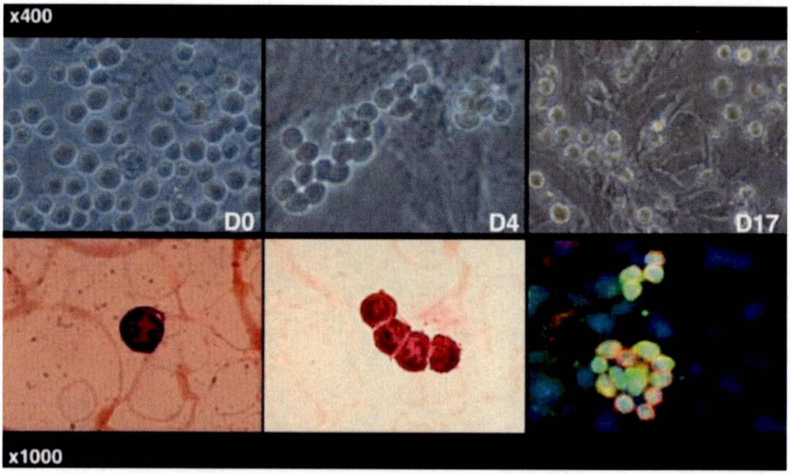

Figure 12. The top row (400x) is brightfield microscopy while the bottom are stained cells using Testsimplet®. After germ cell enrichment, cells were seeded (D0 of culture) while others were stained for morphological analysis. At D4, some germ cells proliferate, and by D17, colonies that had double expression of AP activity and VASA were identified.

At different time points, we assessed the culture to determine the growth of the SSCs (Figure 12). At day 0, unfortunately the TestSimplet® had somewhat dehydrated the germ cells but the large nucleus-cytoplasmic ratio can still be seen. At day 4, we observed some aligned cells, and at day 17, we had some clustering of cells that displayed the AP and VASA stain. At day 21, proliferation was still progressing (Figure 13). Thus, in subsequent experiments we used the medium supplemented with 15% FBS.

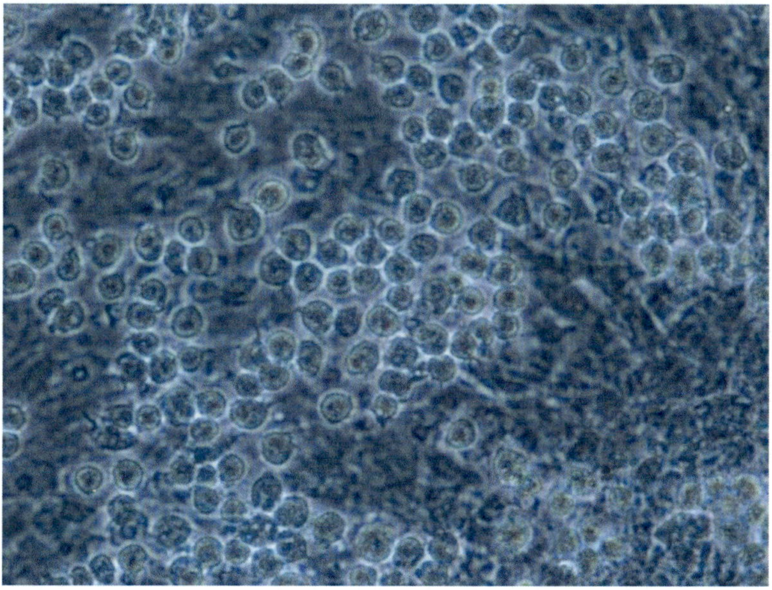

Figure 13. Putative SSC colonies proliferating at day 21.

To test if the feasibility of further promoting SSC growth, the unselected suspensions of testicular cells consisting of a mixture of SSCs, pre-Sertoli cells, and interstitial cells were plated under several culture conditions: 1) GDNF+bFGF or 2) FSH+bFGF or 3) GDNF+bFGF+LIF. Already the following day, epithelial cells had become attached to the dish to form a confluent monolayer with scattered adherent germ cells (Figure 14).

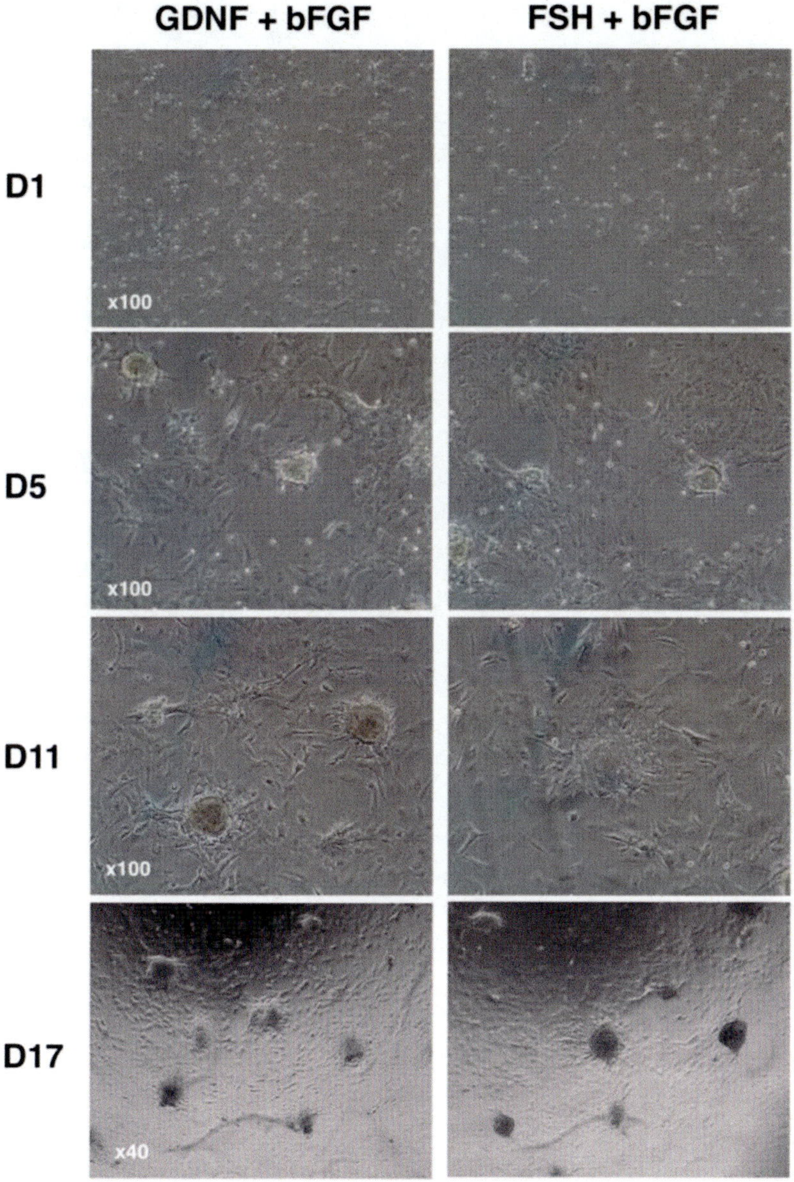

Figure 14. Neonatal testicular cells started to form cell clumps on D1 of culture regardless of the growth factor supplementation combination. Cells began to adhere to the dish and form scattered clusters (D5). These cell aggregates progressively grew over time as seen on frames D11 and D17.

On D3, cell clumps of ~25μm diameter were regularly scattered in the dish. Over the subsequent days, there was a spontaneous and progressive migration of what appeared to be the somatic cells around each germ cells, creating cell clumps at almost regular distribution. In every media combination, the size of the cell aggregates gradually increased to reach a diameter of ~100μm by D10 (Figure 15).

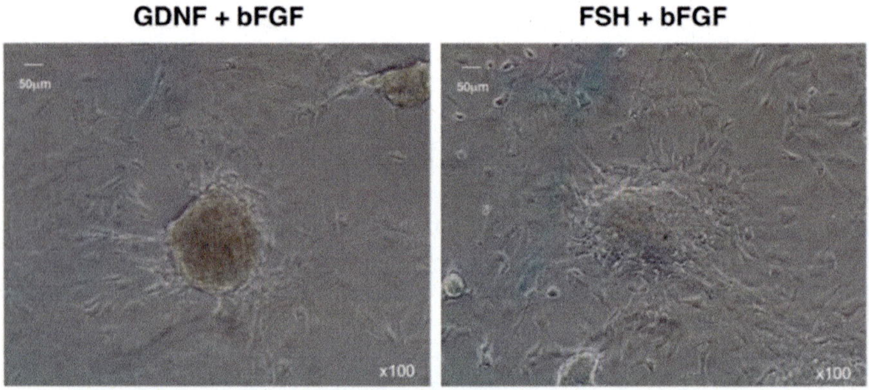

Figure 15. Extended culture of unselected testicular cells supplemented with different growth factors. On D17, in GDNF+βFBF were tightly paced while FSH+βFBF began to spread over the dish.

In GDNF+bFGF and GDNF+bFGF+LIF, the aggregates remained as tightly packed entities, whereas the clumps in FSH+bFGF had started to spread. When LIF was removed, cell aggregates behaved similarly to those in FSH+bFGF. The putative germ cells in these tridimensional seminiferous tubule-like structures were then monitored for up to 75 days. Such cellular aggregates had a steady linear growth up to D56 and then their development plateaued thereafter (Figure 16).

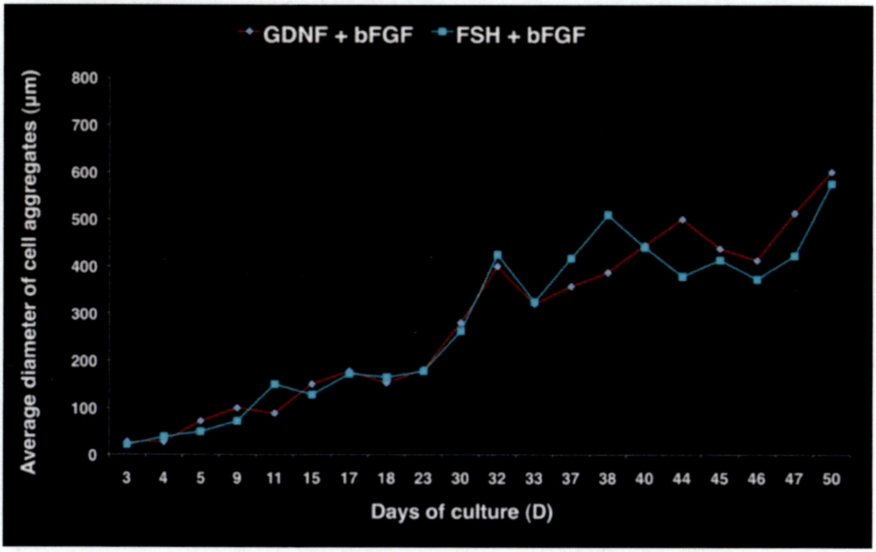

Figure 16. Size of cell aggregates progressively increased with time.

Aggregates were fixed at different times to assess presence of germ cells. Surprisingly, the VASA expression mainly presented on the outer periphery of the aggregates with only seldom individual cells within them, confirming that the germ cells survived the culture for an extended time even if on the outskirt of the cell conglomerate rather than within (Figure 17).

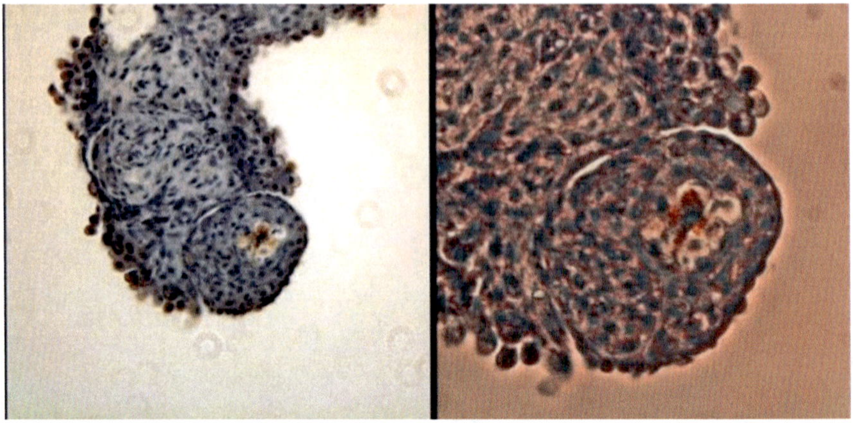

Figure 17. VASA staining (reddish-brown) of testicular aggregates on day 67 of culture were observed in some cells at 100x (left) and 400x (right).

The GDNF+bFGF supplemented with LIF appeared to provide the best support for the germ cell population. Interestingly, after 2 months in culture, VASA staining revealed only a few PGCs within the aggregates, while the majority coating them, presumably SSCs, adopted the form of an inverted seminiferous tubule (as described by Schlatt, ASRM Course 21, 2007). Very few cells positive for FE-J1 (~1%; post-meiotic stage) were observed.

5.5 3D Biomimetic Scaffolding

Another important component to the full establishment of SSC *in vitro* is the incorporation of an appropriate nurturing cell layer (Kanatsu-Shinohara et al. 2003). In the 1960s, Steinberger and associates found in numerous experimental trials that culturing germ cells that were mechanically or enzymatically dissociated from the testis, had their survival directly dependent upon Sertoli cell co-culture (Steinberger et al. 1964; Steinberger et al. 1964; Rassoulzadegan et al. 1993). Recently, differentiation of rat germ cells into spermatids from preleptotene spermatocytes has been documented using Sertoli cells as a feeder layer (Ogawa et al. 2006).

Various approaches to the challenge of *in vitro* spermatogenesis have been attempted, these mainly focusing on testicular cell cultures that are directed at achieving male germ cell differentiation. Among them, tissue culture, organ culture, and co-culture systems have been investigated extensively to assess which system can obtain meiotic or post-meiotic differentiation of these male germ cells. The propagation and maturation of SSC share proliferation challenges and limitations in cell-cell interactions encountered with embryonic stem cell differentiation into germ cells in monolayer systems.

In this section, we utilized a variety of substrates in an attempt to generate a three-dimensional structure aiming at reproducing the seminiferous tubule environment for the isolate SSCs.

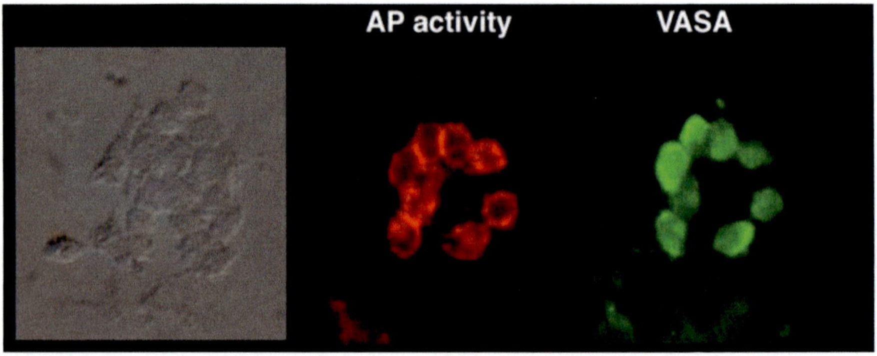

Figure 18. Putative SSC colony growing on mouse embryonic feeders (MEFs) at day 4 of culture (brightfield). The colony expressed both alkaline phosphatase activity (red) and VASA staining (green).

In our preliminary experiments, testes retrieved from seven-6-dpc mice were sorted utilizing the magnetic cell sorting. The magnetic separation yielded an 80% enrichment of SSCs ($P < 0.0001$; see section 5.3). The isolated germ cells were then seeded on a mouse embryonic feeder (MEF) layer to test the ability of this culture system to facilitate germ cells to proliferate. Only one putative SSC colony was identified on day 4 (Figure 18) and proliferated only up to day 9 of culture (Figure 19) (Tanaka et al. 2006).

Figure 19. Putative SSC colony growth up to day 9.

Due to the limited germ cell colony formation, we investigated the effect of different types of substrates in favoring SSC growth *in vitro*. For this experiment, cells were either kept in 1) embedded in collagen matrix at different depths, 2) differential plating utilizing collagen followed by laminin, 3) 0.5% alginate hydrogels, or 4) enclosed in a thin-walled sphere composed of alginate (Neri et al. 2009).

Testes obtained from neonatal mice (6 dpp) were mechanically and enzymatically digested. The enrichment of SSCs were obtained from six neonatal testes along with a small proportion of pre-sertoli and interstitial cells were placed in each culture system to compare which had the ability to propagate and maintain SSCs. Thereafter, the enriched SSCs were placed on 60 mm dishes that had a collagen matrix that consisted of 90% type I collagen with 10% type III, the major components of the basement membrane. Germ cells were resuspended in the collagen matrix and allowed to polymerize where the matrix was approximately 3mm thick (Figure 20a). On day 1, very few SSCs were located in the matrix, which could be due to the fact of its opacity and thickness (Figure 20b). Several days later, very few spermatogonia were dividing (Figure 20c).

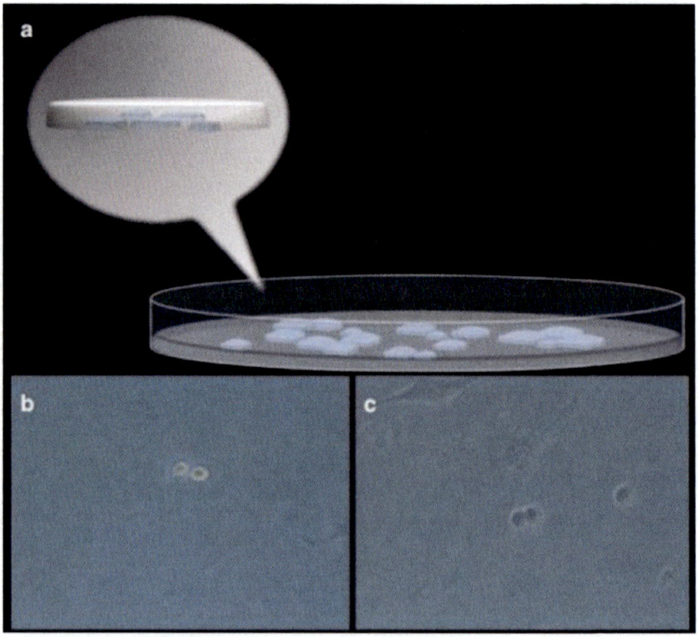

Figure 20. Collagen plating. SSCs were seeded into a collagen mixture (a). A couple of germs cells were found on day 1 (b) and a limited number actually proliferated on day 4 of culture (c).

In the following comparison, we utilized a similar collagen mixture but seeded the germ cells right after collagen polymerrization, leaving them only partially embedded (Figure

21a). Only a limited number of germinating cells were found on day 1 (Figure 21b). In fact, during medium changes tremendous amounts of dead cells were removed leaving only some rare individual SSCs within the matrix (Figure 21c).

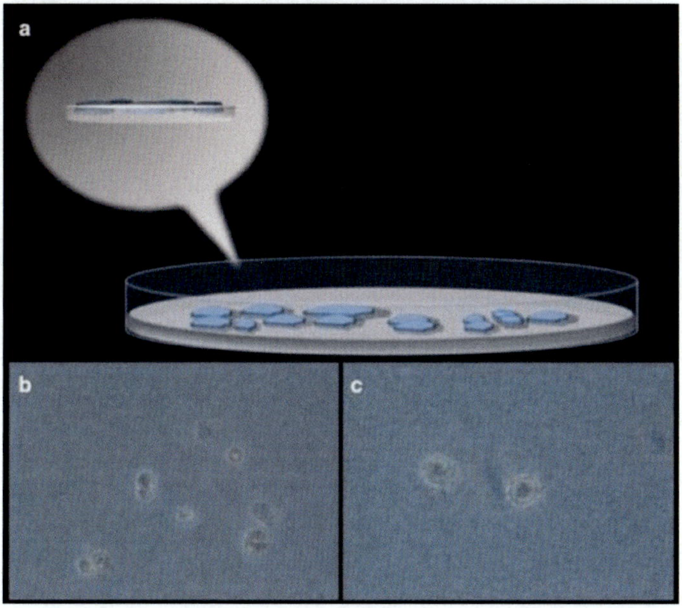

Figure 21. Collagen plating. SSCs were seeded into a collagen mixture post-polymerization (a). A couple of germs cells were found on day 1 (b) and very few on day 4 of culture (c).

In the third model, culture dishes were thinly coated with the 90% type I + 10% type III collagen mixture followed by SSC layering on top (Figure 22a). After se-eding, we we-re able to clearly see the location of each germ cell (Figure 22b). By day 5, colonies began forming (Figure 22c) and growing in size as seen here at day 9 (Figure 22d) and doubling in number as seen in day 14 (Figure 22e).

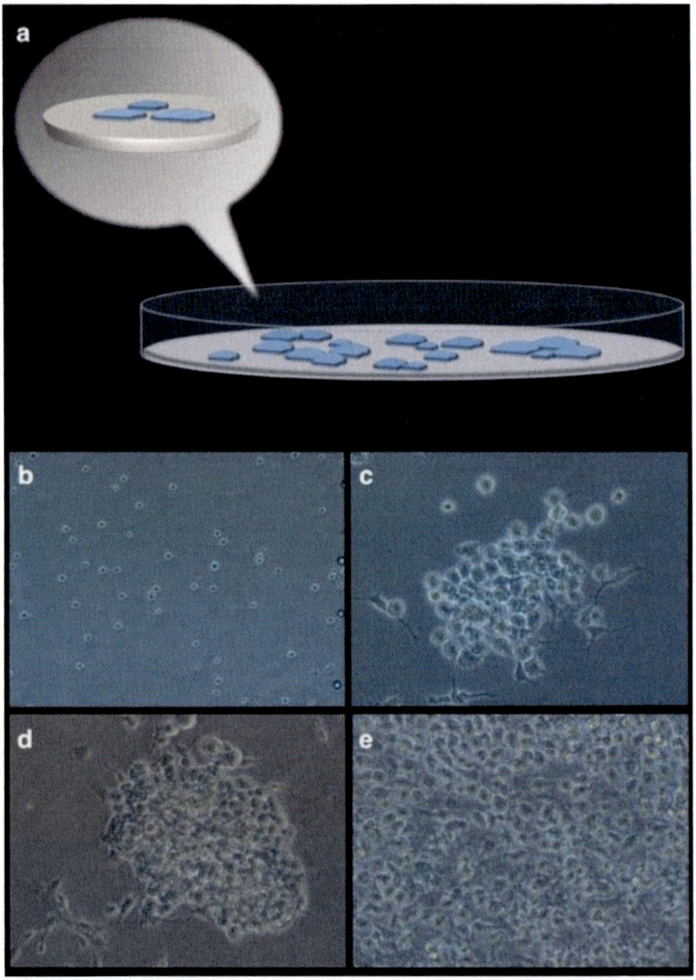

Figure 22. Collagen plating. SSCs were seeded into a thinly polymerized collagen mixture (a). Several individual germs cells were found on day 0 after seeding (b). Colony formation began (c) and continuously growing in size (d,e).

The basement membrane consists of thin layers of extracellular matrix components such as laminin, collagen type IV, and fibronectin, among others. The SSCs were

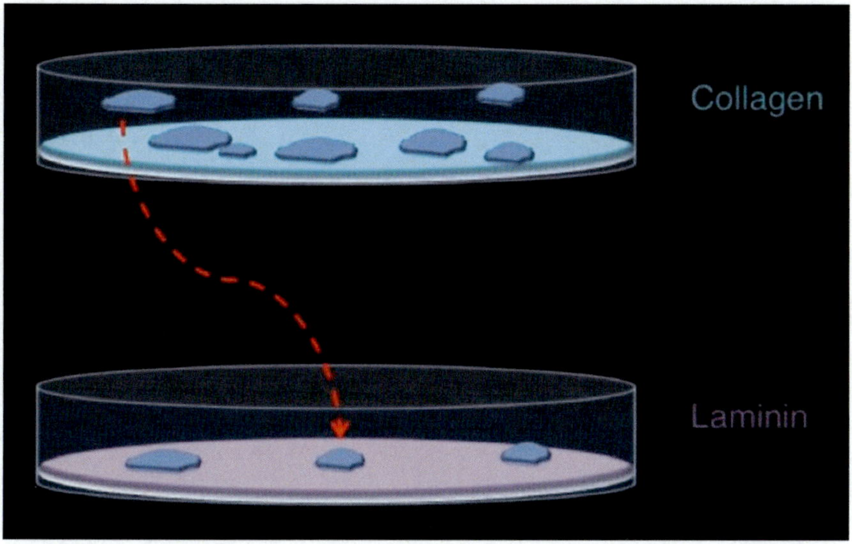

Figure 23. Differential plating utilizing pre-coated collagen culture dish followed by laminin.

observed to show highest affinity to laminin (Shinohara et al. 1999; Orwig et al. 2002). Some differentiated cells, including SSCs, have β_1- and α_6-integrins that take part in binding to laminin (Hynes 1992; Siu and Cheng 2004, 2004).

In a later experiment, we performed differential plating utilizing commercially available dishes such a pre-coated collagen and laminin (Neri et al. 2012). Testicular cell suspension was first placed in the pre-coated collagen for 4 hours (Figure 23). Thereafter, only the floating cells were transferred over to the laminin dish. Those that attached to the collagen were considered to be other testicular cells (Figure 24a), while those on the laminin were presumed to be spermatogonia stem cells (Figure 24b) because of the characteristic tendency to migrate towards the basal epithelium. At day 3, in the laminin coated dish the germ cells were loosely attached and few in number (Figure 24c). At day 6, we saw a slight increase in germ cells, however, there were twice as many floating dead cells as well (Figure 24d).

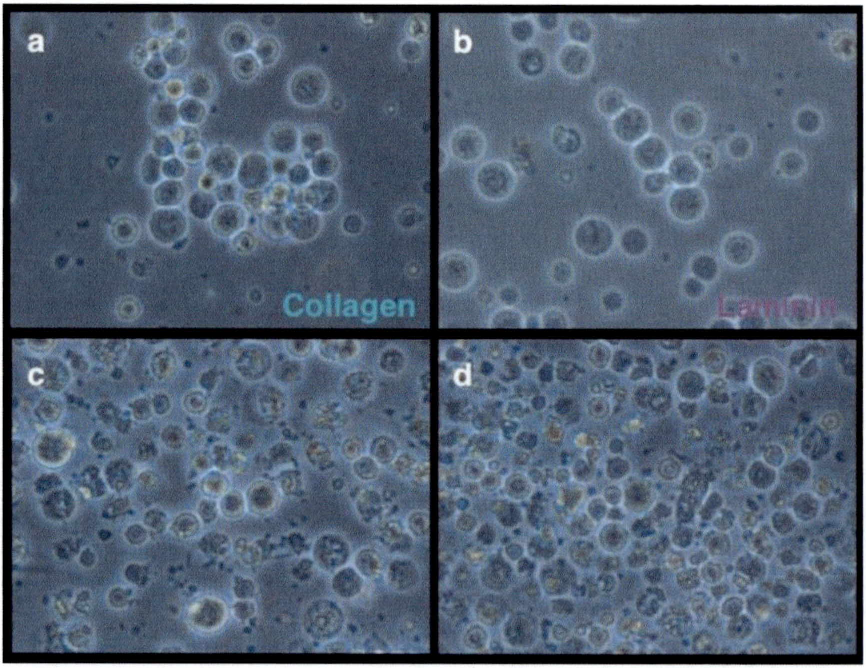

Figure 24. Differential plating where testicular cells that attached to the collagen (a), and SSCs attach to laminin (b). Propagating germ cells on day 3 (c) and day 6 (d).

Since these various substrates were not producing the results we wanted, we decided to conduct a comparative study between adult and neonatal testicular tissue to obtain a spatial orientation of where and how the spermatogenic cells are architecturally and dynamically located during testicular development and maturation. To perform this, mouse testes were fixed and embedded in paraffin. Thereafter, 5 μm sections were stained with VASA (primordial germ cell to early spermatid), DAZL (A spermatogonia), vimentin (Sertoli cells), and 3-β-hydroxysteriod dehydrogenase (3-β-HSD; Leydig cells) (Neri et al. 2009).

After staining with VASA we were able to identify different stages of the germ cells in the adult and neonate, from type A adjacent to the basal membrane followed by type B more centripetally towards the lumen (Figure 25a,b). Spermatocytes in different developmental stages matured and migrated in the same direction. Early spermatids (round) found closer to the lumen were also positive for VASA. When specimens were stained with DAZL (Figure 25c,d), a ring of type A spermatogonia was located next to the basement membrane. Sertoli cells (Figure 25e,f) appeared as finger-like projections going towards the lumen with less definition in the neonate. Leydig (Figure 25g,h) in between the external tubular walls were tightly packed in the adult while fewer were visible in the 6 day old pups.

From these results, in the following experiments we then attempted to generate a three-dimensional structure reproducing the seminiferous tubule environment. We enclosed the germ cell suspension in a semi-permeable membrane utilizing alginate or by applying the principle of reverse spherification (RS).

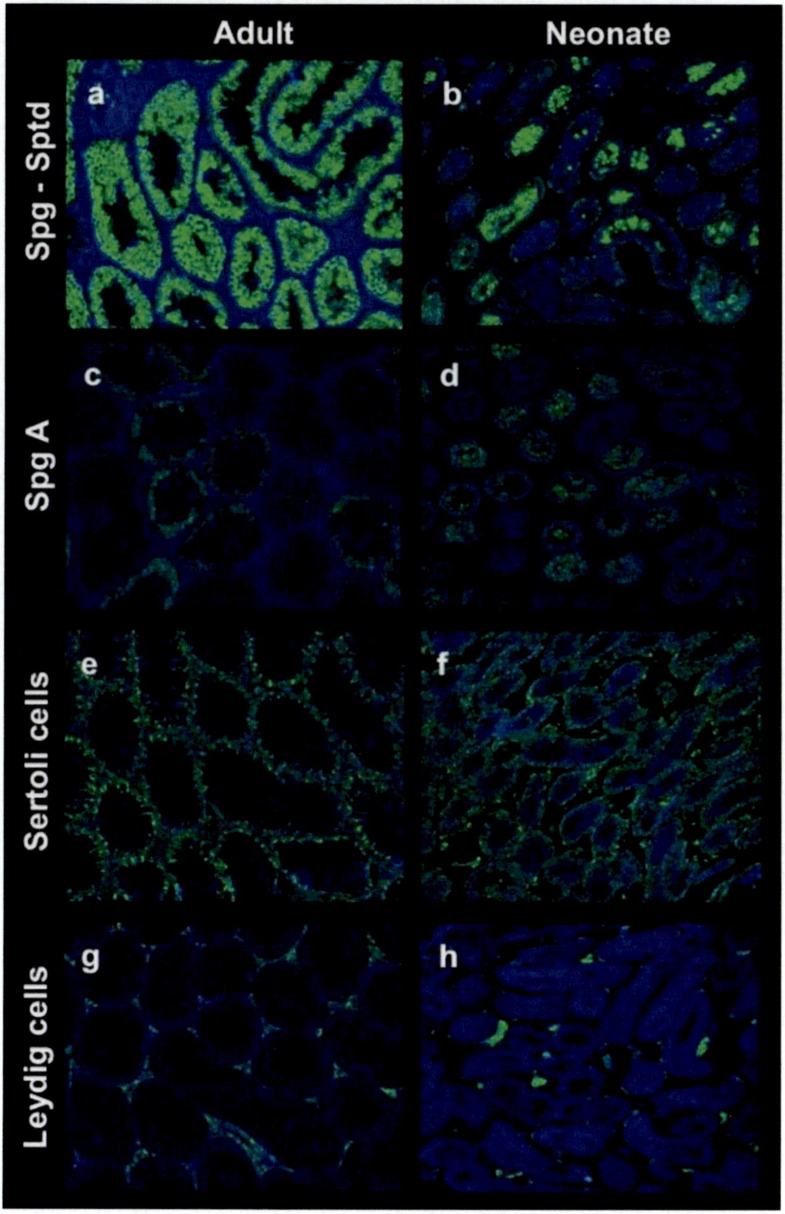

Figure 25. Mouse testes stained with VASA (a,b) to identify spermatogenic cells all the way to early spermatids, DAZL (c,d) for type A spermatogonia, vimentin (e,f) for Sertoli cells, and 3-β-HSD (g,h) for Leydig cells.

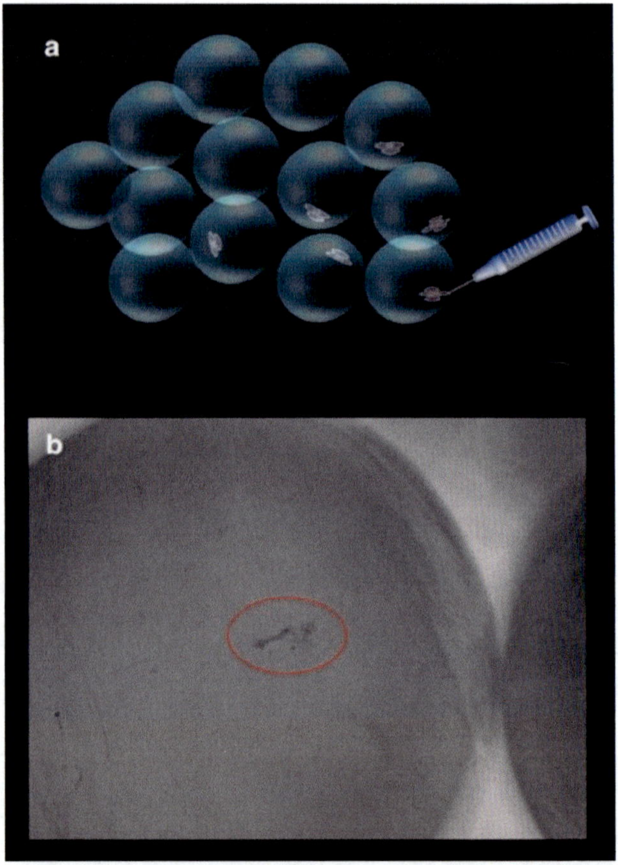

Figure 26. Alginate hydrogels. Beads were injected with enriched SSCs (a). After culture, cells did not proliferate and became apoptotic (red circle) (b).

In the following model, we utilized 2.5% alginate hydrogels. Here, we created beads of 10 μl in size that we injected with spermatogonia cell suspension prior to polymerization of the alginate (Figure 26a). Small droplets were made through a 25G needle (short length) onto the nylon mesh (1" x 1") inside the lid of a 60mm dish. Loading the alginate proved to be extremely difficult. After all beads have been injected with SSCs, the nylon mesh was turned upside down over the 50mM $CaCl_2$/140mM NaCl solution for 30 seconds to be certain that the bead is submerged in the solution to allow for proper cross-linking. Thereafter, the beads were transferred into medium with growth factor supplementation. When excess medium was injected with the cells, the alginate does not polymerize well, either flattening in the shape of a pancake or a segment of the aliginate was lost which means loss of the injected germ cells. Unfortunately, even at 1% alginate composition the germ cells became apoptotic after 3-4 days (Figure 26b). It appears that alginate, just like in the 3 mm collagen matrix, trapped the germ cells preventing them from thriving and migrating.

In our final *in vitro* platform, the reverse spherification model (Figure 27), where cells were enclosed in a malleable membrane germ cell number doubled by day 5 and continued to proliferate up to day 30 as confirmed by molecular markers. Moreover, in about 10% of the spheres, we were able to identify cells that clearly reached the post-meiotic stage.

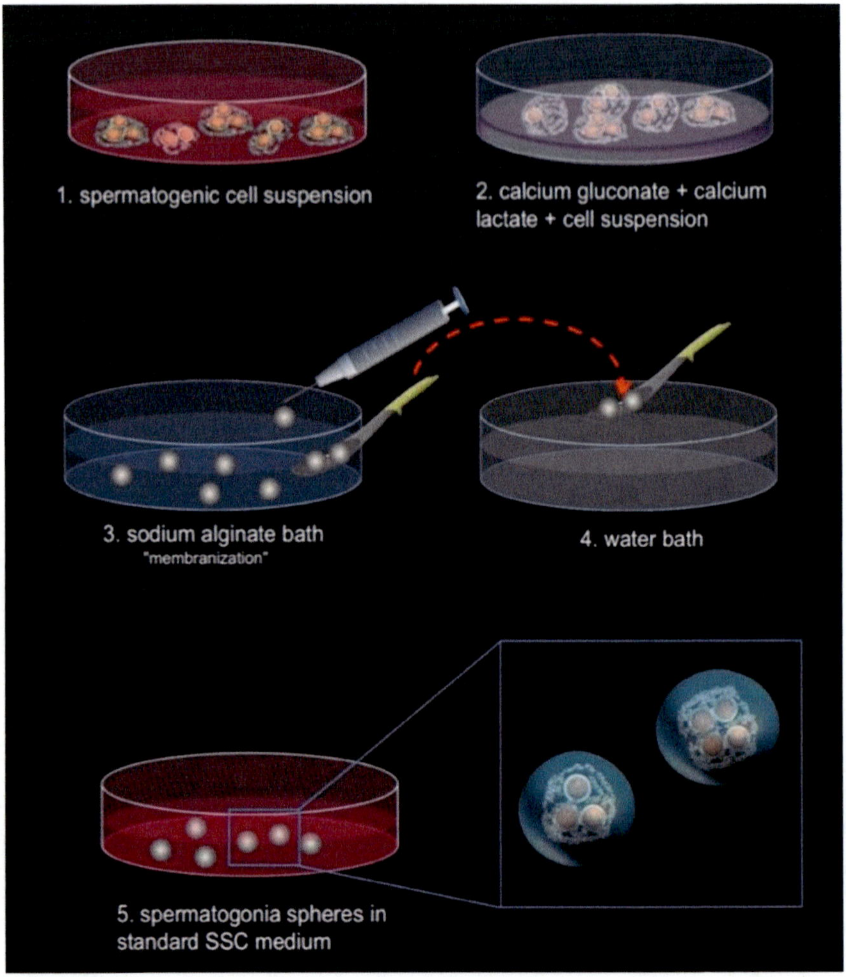

Figure 27. Neonatal spermatogenic cells in culture were placed in a 1:1 ratixo of calcium gluconate/lactate bath. Thereafter, cells were "membranized" by placing small drops in the alginate bath. Excess alginate is removed in the waterbath and spermatogonia spheres are transferred into our standard SSC culture medium.

5.6 Characterization of Immature Human Germ Cells

5.6.1 Men with Secretory Azoospermia

Azoospermia affects about 10% of all infertile men. While obstructed azoospermic men had their spermatogenesis preserved and spermatozoa can be easily retrieved either through the epididymis or testes, non-obstructive azoospermic men on the other hand, yields spermatozoa in only 60% of the biopsies (Schlegel 2013). These men with testicular failure either have hypospermatogenesis in about 50% of the cases (Figure 28a), while 25% of them present with some level of maturation arrest where only immature germ cells are found in

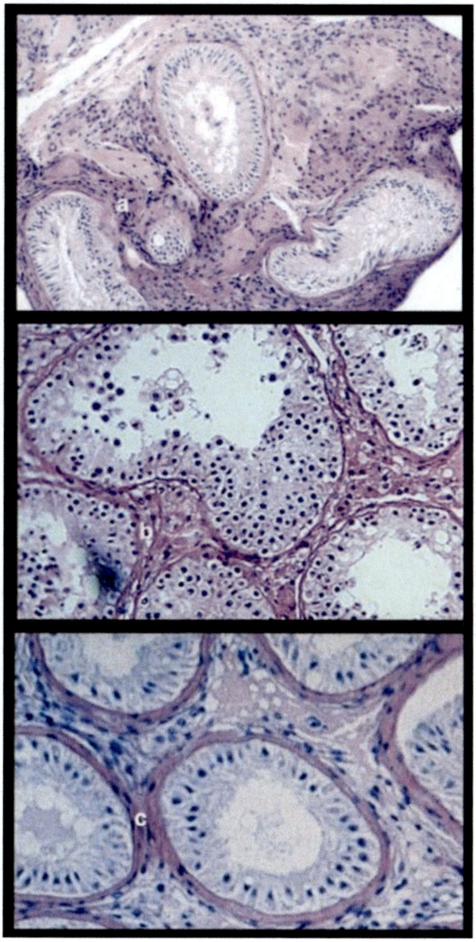

Figure 28. Testicular biopsies of men diagnosed as hypospermatogenesis (a), maturation arrest (b), and Sertoli cell only syndrome (c).

their surgical specimen (Figure 28b), and the remainder of men with secretory azoospermia have germ cell aplasia also known as Sertoli-cell-only syndrome (Figure 28c).

In consideration of the fact that the abovementioned diagnoses are based on histological findings of a randomly selected biopsy specimen, these histological pictures do not always represent the actual condition throughout the testicle and, therefore, do not predict the successful retrieval of fully developed spermatozoa and vice versa. Thus, testicular biopsies performed on azoospermic men scheduled for ART treatment with ICSI at times fail to yield spermatozoa. In these instances, we investigated

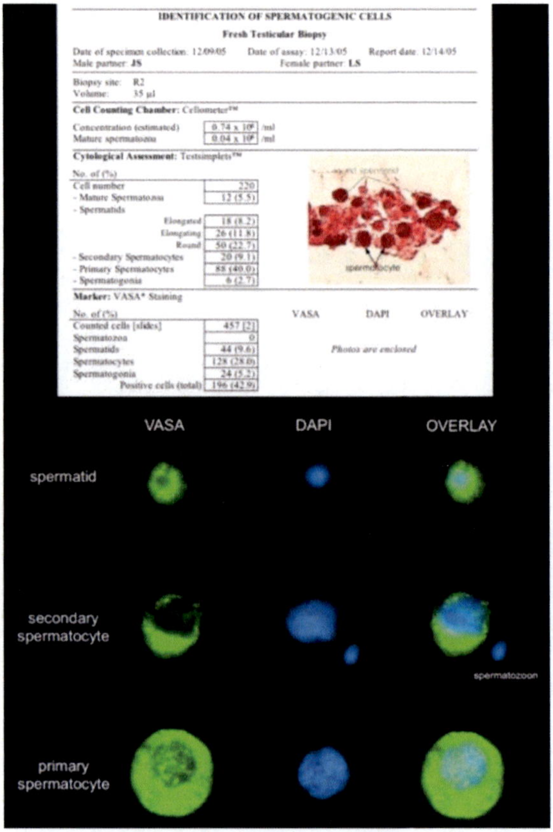

Figure 29. Identification of spermatogenic cells in an NOA man undergoing ICSI.

the feasibility of assessing the residual germ cells in digested cell suspensions and seminiferous tubules fragments.

For this purpose, a total of 10 testicular biopsies from NOA men were processed, of which 5 were post-enzymatic digestion and 3 included seminiferous tubular fragments of 2 and 5mm. In the first case, after testicular biopsy retrieval samples were digested and processed for ICSI. A total of 9 oocytes were retrieved and injected with the motile testicular specimen. The remainder of the sample was processed with VASA to identify the different spermatogenic cells present in the sample (Figure 29). A total of 0.04 x 10^6/ml of spermatozoa was identified together with approximately 740,000 round cells. After staining, we were able to identify 9.6% of the cells to be spermatid, 28.0% spermatocytes, and 5.2% spermatogonia (Figure 29).

In another instance, we wondered if we missed spermatozoa in a specimen following an exhaustive sperm search of over 4 hours carried out by 8 embryologists. A portion of the specimen was stained for morphometric characterization of germ cells and VASA expression. After counting over 200 round cells, we were only able to identify one that was stainable and had the appropriate nucleus/cytoplasmic ratio.

We then analyzed other 8 men who underwent testicular biopsy at our Center. Testicular samples were provided by the same reproductive urologist and processed (Figure 30a-c).

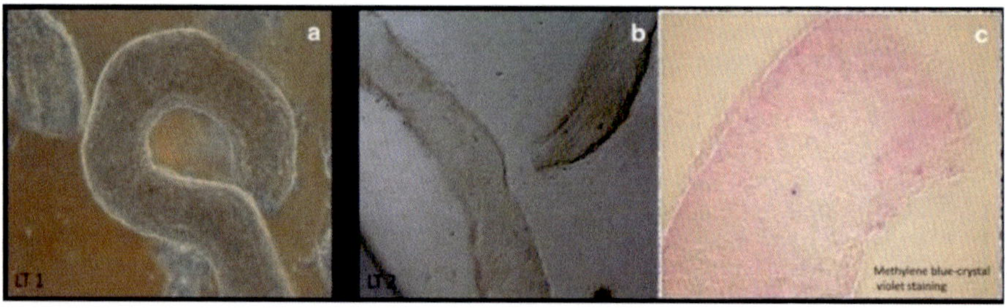

Figure 30. Human testicular tubules pre-digestion (a), post-digestion (b), and after simple staining (c).

In only 2 men were we able to find germ cells in the digested cell suspensions, where a total of 606 cells were counted consisting of 25 (4.1%) spermatogonia, 128 (21.2%) spermatocytes, and 44 (7.3%) spermatids (**Table 6**).

Table 6. Identification of germ cells in NOA men undergoing the ICSI procedure

No. of (%)	Specimen 1	Specimen 2
Biopsies analyzed	8	
Germ cells (+)	2 (25.0)	2 (25.0)
Spermatogonia	11/268 (4.1)	14/338 (4.1)
Spermatocytes	68/268 (25.4)	60/338 (17.9)
Spermatids	—	44/338 (13.0)

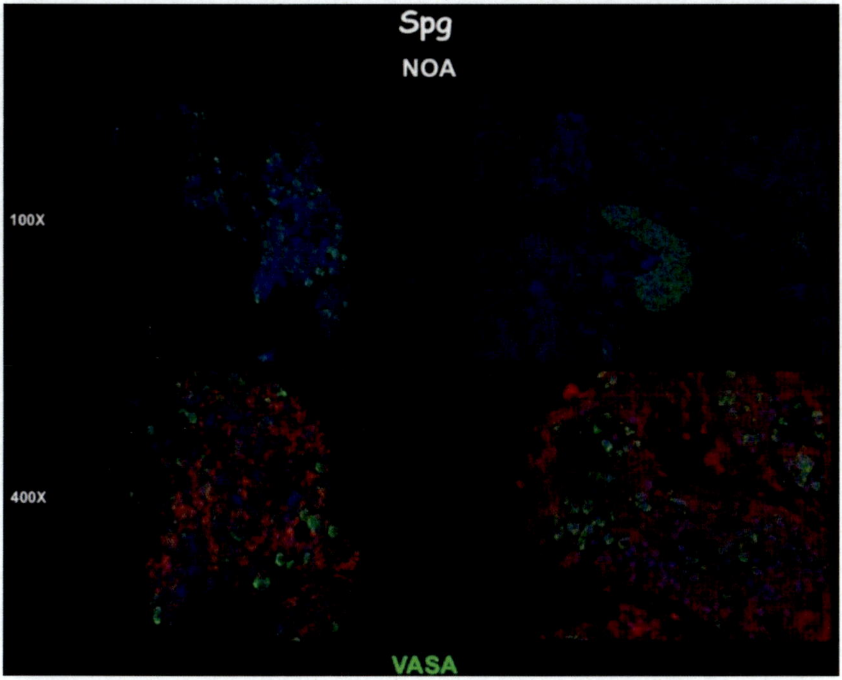

Figure 31. Tubules from NOA samples, green indicates VASA positive cells and blue is nuclear counterstain.

When intact tubules were evaluated, germ cells were identified by VASA in these NOA men. However, there were very few tubules with spermatogonia in them (Figure 31).

Due to the mincing of testicular tubules in the attempt to release spermatozoa, the typical seminiferous tubule architecture has been lost, hence the Sertoli (Figure 32a) and Leydig cells (Figure 32b) were not well demarcated and randomly distributed (Daw et al. 2011).

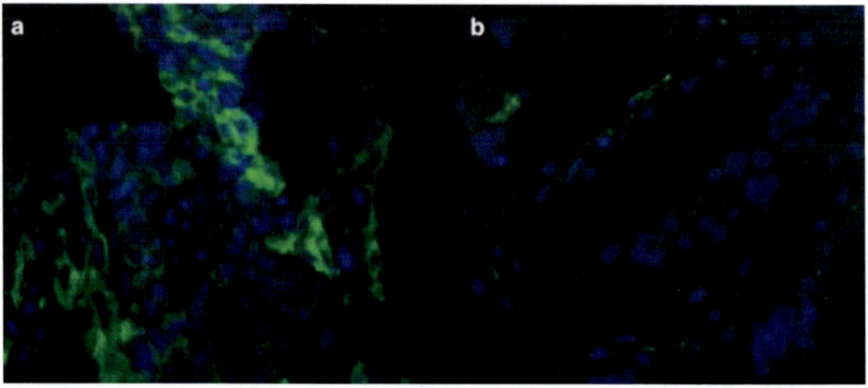

Figure 32. Minced human testicular biopsies stained for vimentin (a) and estrogen receptors (b).

A significant limitation of testicular biopsies and even micro-TESE is the inability to identify seminiferous tubules that will actually contain intact germinal epithelium and therefore spermatozoa. Sperm containing tubules are evaluated by subjective assessment of tubular size and appearance. With this approach, spermatozoa have been retrieved in 40% to 60% of men depending on the underlying cause of their hypospermatogenesis (Okada et al. 2002; Ramasamy et al. 2005). The introduction in the field of reproductive urology of the multiphoton microscopy (MPM) enables label-free immediate visualization of many biological processes in living tissue at subcellular resolution. In collaboration with our urology colleagues, we have used the MPM to directly visualize rodent tubules to improve the success of micro-TESE. Here, a custom built system was used for MPM imaging (Figure 33) (Ramasamy et al. 2011).

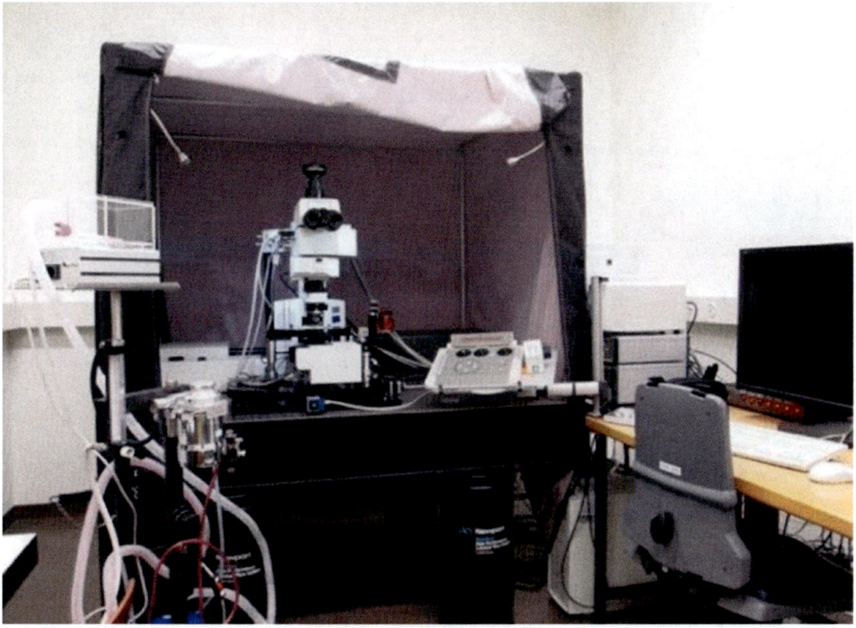

Figure 33. Multiphoton microscopy set-up.

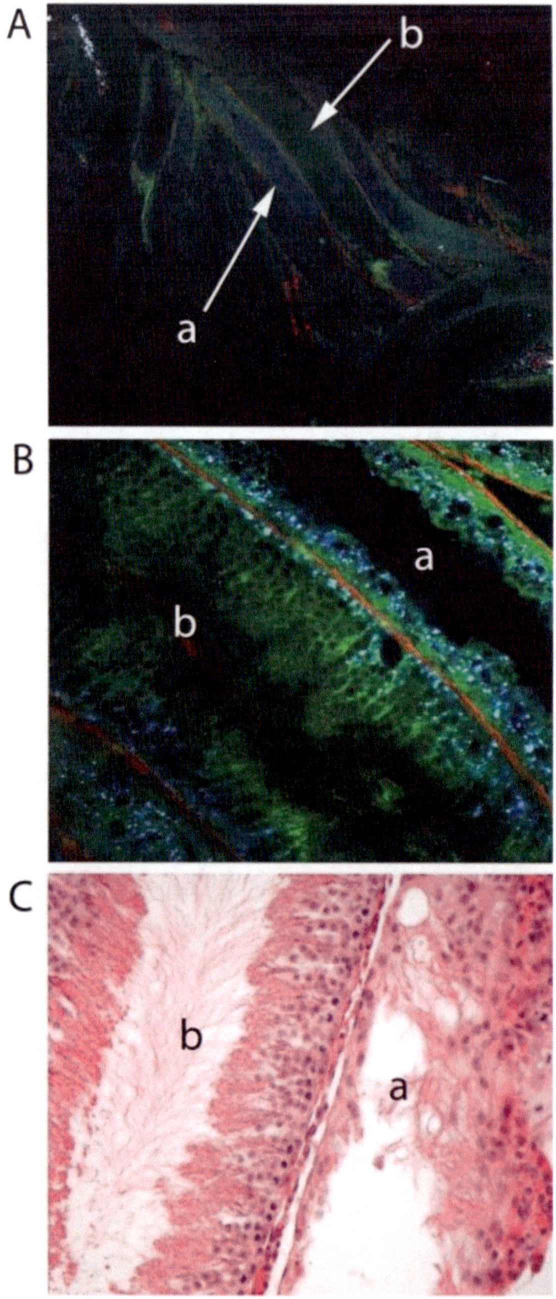

Testes was placed at the center of a glass bottomed 35 mm dish with the seminiferous tubules facing up, hydrated with PBS and overlaid with a coverslip. A drop of PBS was placed on the coverslip to achieve objective water immersion. The whole dish was placed under the desired objective of the

Figure 34. Multiphoton microscopy imaging of rodent seminiferous tubules. Germ cell depleted, Sertoli cell only seminiferous tubule (a) and tubules with spermatogenesis (b). Low power MPM (A). Reduced from x4. High power MPM (B). Reduced from x25. H&E, reduced from x25 (C).

upright microscope. The typical interval between orchiectomy and data acquisition was less than 30 minutes. We used the MPM to evaluate the various histopathological conditions such Sertoli-cell-only and maturation arrest in rodents. These histopathological conditions replicate those likely to be found in infertile men. In this sample, a heterogenous appearance of tubules recapitulated the full spectrum of testicular histopathology in infertile men (Figure 34a-c). Tubules with spermatogenesis versus Sertoli-cell-only tubules were easily distinguished by MPM scanning at low and high magnification.

5.6.2 Xenogeneic transplantation

Successful spermatogonia transplantation has been reported in the mouse, where germ cells from a fertile mouse were transplanted into the seminiferous tubules of an aspermatogenic one (Brinster and Avarbock 1994; Brinster and Zimmermann 1994). Similarly, reinitiation of spermatogenesis has been observed when rat germ cells were transplanted into immunodeficient mice. This suggests that spermatogenic stem cells of different species may be successfully transplanted (Clouthier et al. 1996), offering the possibility of xenogeneic spermatogenesis of human spermatogonia.

Human male germ cell transplantation may be a valuable technique to obtain fully developed spermatozoa and when used in conjunction with ICSI may offer an alternative treatment for those patients with spermatogenic arrest. Furthermore, it would provide an innovative way to preserve fertility in oncological patients undergoing chemo-radiotherapy, with particular benefit for pre-pubertal cancer patients.

In this study, we attempted to restore a human spermatogenetic line in testes of mutant aspermatogenic recipient mice (W/W^v) and immunologically anergic mice (SCID) transplanted with germ cells isolated from azoospermic men. The feasibility of injecting mouse seminiferous tubules using a simple manual control technique was investigated, with a focus on the efficiency of the injection procedure.

In our first attempt, we digested a testicular biopsy obtained from an obstructive patient with polycystic kidney disease and multiple epididymal cysts. Isolated testicular spermatogenic cells were stained for flow cytometric analysis in an attempt to separate the different ploidy status of the germ cells (Figure 35).

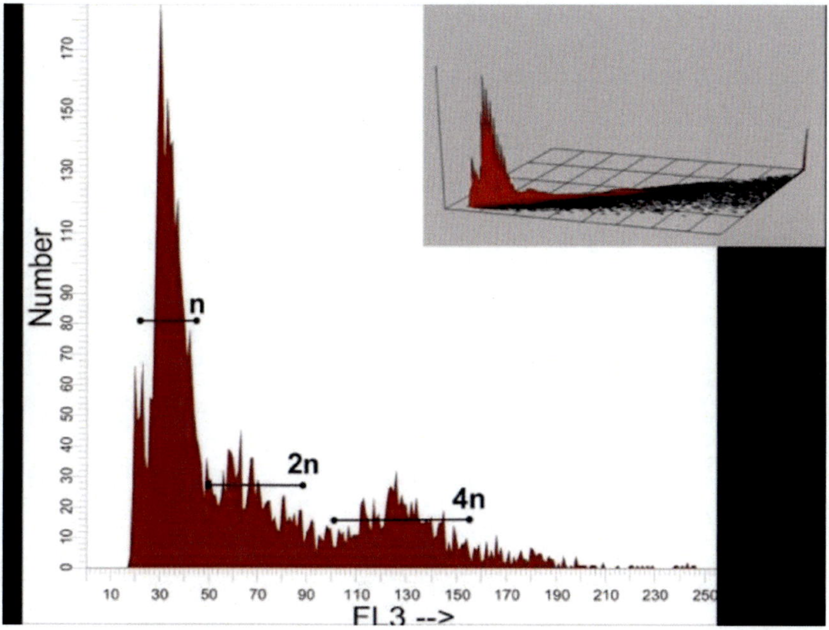

Figure 35. Flow sort histogram result of an OA sample indicating the population that was haploid (n), diploid (2n), and tetraploid (4n).

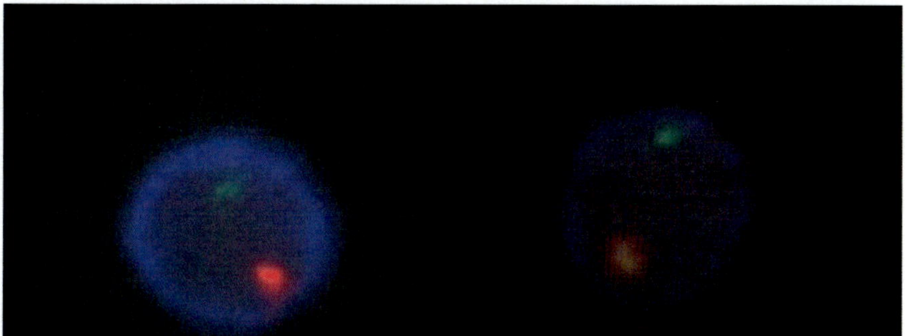

Figure 36. Fluorescent in situ hybridization (FISH) of germ cells indicating haploid cells were identified, Y (red), X (yellow), and 18 (green).

The ploidy of the cells was confirmed by cytogenetic analysis. Due to the limited number of cell obtained, the sample was not utilized for injection. Approximately 45,000 cells were sorted from an initial group of 100,000. The incidence of haploid cells was 14.0% (*n* = 6,200), diploid was 43.1% (*n* = 19,400), and tetraploid was 26.0% (*n* = 11,700) confirming the hypospermatogenesis. To confirm ploidy and cell sorting efficiency, FISH was performed (*n* = 23), with the finding of 26.1% (6/23) Y18, 21.7% (5/23) X18 for the haploid cell group, 30.4% (7/23) diploid, and 21.7% (5/23) aneuploid cells (Figure 36).

The first xenogeneic germ cell transplantation in which human testicular cells were injected into mouse seminiferous tubules was performed by our group in 2000 (Reis et al. 2000) that resulted in a lack of recolonization of SSCs in the recipient tubules of SCID or W/W^v mice after 150 days.

Spermatogenic cells were obtained from testicular biopsy specimens of men (average age 34.3 ± 9 years) undergoing infertility treatment because of obstructive and non-obstructive azoospermia. Testicular tissue was digested with collagenase to promote separation of individual spermatogenic cells. The germ cells were injected into testicular seminiferous tubules using a microneedle with a 40 μm inner diameter on a 10 ml syringe (Figure 37). To assess the penetration of the cell suspension into the tubules, trypan-blue was used as an indicator. Mice were maintained from 50 to 150 days to allow time for germ cell colonization and development prior to sacrifice. Then, testes were fixed for histological evaluation and approximately 100 cross-sectioned tubules were examined for human spermatogenic cells. A total of 14 testicular cell samples,

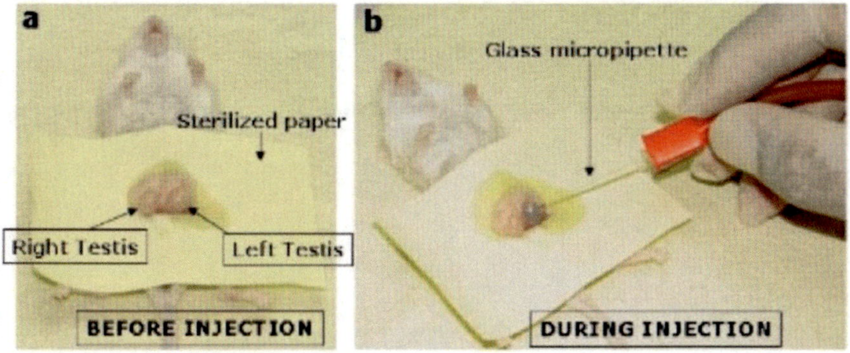

Figure 37. Xenogeneic transplantation of spermatogonia stem cells. A mouse is draped and testicles excised (a). Injection of SSC (with trypan blue) into the left testis (b).

8 frozen and 6 fresh, were obtained from 12 men. The origin of the azoospermia was obstructive for half of the patients and NOA for the remainder. The concentration of spermatogenic cells in the OA group was 9.4 x 10^6, and for the NOA group was

1.6×10^6 ($P = 0.01$). The different spermatogenic cell types were distributed equally in the OA samples, ranging from spermatogonia to fully developed spermatozoa, but in the NOA group the majority of the cells were spermatogonia and spermatocytes. A total of 23 testes of 14 mice were injected successfully, as judged by the presence of spermatogenic cells in histological sections of testes removed immediately after the injection. However, sections from remaining testes examined up to 150 days after injection showed tubules lined only by Sertoli cells, but without any xenogeneic germ cells (Figure 38).

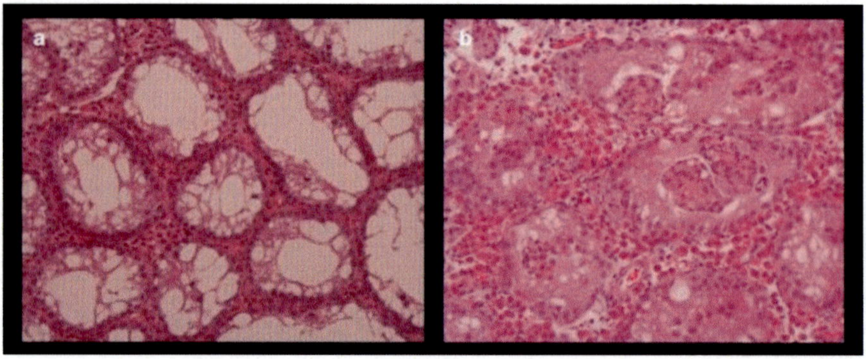

Figure 38. Seminiferous tubules of SCID mouse 4 weeks after busulfan treatment (a) and 1 week after injection of human spermatogenic cells (b).

We have postulated then that lack of success may have been due to inter-species variability, to non-compatible cell adhesion molecules and/or to immunological rejection.

Nagano and colleagues transplanted baboon germ cells into mice and found that germ cells were able to repopulate and survive for 6 months (Nagano et al. 2001).

Investigators have reported the survival of at least some undifferentiated spermatogonia during distant xeno-germ cell transplantations (Reis et al. 2000; Nagano et al. 2001; Nagano et al. 2002). In Reis et al., however, we did not find evidence of donor tissue survival following a human-to-immunodeficient mouse testicular tissue transplantation. In this study, the antibody stain for proacrosin was used to detect successful tubular implantation. Proacrosin is a marker of differentiated human spermatogonia (primary spermatocytes and spermatids) and would not have detected transplanted cells that had survived or propagated without further differentiation. On the other hand, Sofikitis et al reported successful spermatogenesis from transplantation of human tissue into rat and mouse recipients (Sofikitis et al. 1999). In 2002, Nagano and colleagues reported on the use of anti-baboon testes antibody to identify the survival of human spermatogonia in mouse recipients for up to 6 months post-transplantation (Nagano et al. 2002). The same group then used human SSCs but found that these cells had not differentiated beyond spermatogonia 1 month after transplantation to the mouse testes. These interspecies transplantation studies seem to suggest, at present, that the vast phylogenetic distance between species (baboon to mouse and human to mouse) will likely translate into host testicular environment and donor spermatogonia incompatibilities that prohibit completion of spermatogenesis.

Chapter 6

Conclusions and Future Perspectives

ICSI is the most effective treatment currently available for severe male factor infertility and requires fully developed spermatozoa but is ineffective for men suffering from complete spermatogenic arrest. In mammals, millions of spermatozoa are produced daily, ultimately from spermatogonia stem cells. These precursors of spermatogonia depend on specific growth factors for survival and proliferation but also on the close relationship with the enveloping presumptive Sertoli cells which seems to prevent apoptosis of these germ cells. The ability to propagate and immortalize SSCs in vitro would allow the creation of colonies, which on intra-testicular transplantation might be able to repopulate the germinal epithelium of azoospermic men with Sertoli-cell-only syndrome.

Men with germ cell aplasia as well as men and pre-pubescent children surviving cancer who had undergone chemotherapy treatment require the transplantation of germ cells to re-seed the spermatogenic process. Since chemotherapy involves the use of extremely cytotoxic chemicals, it can cause tremendous harm to germ cells, often rendering the patient sterile.

In this regard, significant strides have been accomplished in experimental animals through spermatogonia transplantation that offers an opportunity for future clinical application in humans. A handful of researchers have been able to recolonize the basal membrane of seminiferous tubules by injecting SSCs, therefore re-establishing spermatogenesis in autologous transfers. The so derived viable spermatozoa were capable of fertilizing mature oocytes resulting in the production of progeny. As experiments involving human subjects are limited by ethical constraints, xenogeneic transplantation using experimental models has helped to provide an adequate environment to study the different stages of germ cell differentiation. Attempts have been made at Cornell to transplant surgically retrieved human germ cell suspensions into the rete testes of immunodeficient mice, but unfortunately without success. However, in our study the choice of the post-meiotic marker may have been a little ambitious and therefore an earlier germinal stage may have actually proliferated within the transplanted tubule and may have been missed. These experiments appear to be plagued by proliferation limited to premeiotic stages and have no maturational capacity towards the completion of spermatogenesis.

At the present time, cryopreservation methods are limited to utilizing standard freezing methods either with DMSO or glycerol. Another method that has been attempted to assess if spermatogonia cells have better survival rates is vitrification. However, the main limitation of cryopreservation is the possible damage to the spermatogonia thereby limiting its survival as well as its ability to self-renew. In this instance, the addition of growth factors may help germ cell survival, proliferation, and eventual differentiation to mature gametes.

The development of long-term ex vivo maintenance of testis tissue utilizing microfluidic devices may help overcome present day obstacles and improve culture conditions for various tissues and organs.

With the availability of embryonic stem cells the possibility to recreate a male germ cell niche, through the tridimensional model represented by the embryoid bodies and three-dimensional scaffolding, appears to be concrete. While these studies indicate that alternative sources of gametes are not merely the result of science fiction but a reality, further advances on the epigenesis of the neo-gametes are welcome. Moreover, the possibility to propagate a male genome would provide an alternative means through which to consistently obtain conceptuses.

Once the abovementioned steps have been achieved reproductive medicine will forever be changed, allowing scientists and physicians to develop a feasible treatment that will restore fecundity in a sterile man. Thus, spermatogonia stem cells and their subsequent *in vitro* culture is the future of reproductive medicine, finally providing chances to infertile couples that wish for nothing more than starting a family of their own.

Acknowledgments

I express my gratitude to my parents and my loving brothers, for their unselfish dedication and support throughout my life and career.

I wish to thank Gianpiero Palermo for stimulating my interest and transmitting the dedication to Reproductive Biology and Medicine, and for always being there when I needed him.

A special thanks goes to Alan Transon and David De Kretser for allowing me to be their student and providing me with such a unique opportunity.

I wish to thank Grahm Jenkin for allowing my PhD dream to become a reality. I will always be grateful for that.

I am very appreciative to all clinicians and scientists at the Center for Reproductive Medicine of Weill Cornell Medicine for the assistance with this study. I am grateful to all the research associates and fellows that I had the chance to work with, Takumi Takeuchi, Yukiko Katagiri, and Noriko Tanaka. Without their support, help, and dedication, this work would not have been possible.

Chapter 6

I am thankful to the Andrology and ICSI Teams for the support I received from all the members, particularly Trina Fields, Stephanie Cheung, Christopher Chen, Bong Kyo Seo, Tyler Cozzubbo, and Nigel Pereira.

A special thanks goes to Zev Rosenwaks for giving me the great opportunity to grow in his Center. He has always been supportive of my career and very fond of my endeavors.

Bibliography

Allen, C., Keane, D. and Harrison, R. F. (2003) A survey of Irish consultants regarding awareness of sperm freezing and assisted reproduction. *Ir Med J*, **96**, 23-25.

Baniel, J. and Sella, A. (2001) Sperm extraction at orchiectomy for testis cancer. *Fertil Steril*, **75**, 260-262.

Berthelsen, J. G. and Skakkebaek, N. E. (1983) Gonadal function in men with testis cancer. *Fertil Steril*, **39**, 68-75.

Bleyer, A. and Barr, R. (2009) Cancer in young adults 20 to 39 years of age: overview. *Semin Oncol*, **36**, 194-206.

Bleyer, W. A. (1990) The impact of childhood cancer on the United States and the world. *CA Cancer J Clin*, **40**, 355-367.

Brandriff, B. F., Meistrich, M. L., Gordon, L. A., Carrano, A. V. and Liang, J. C. (1994) Chromosomal damage in sperm of patients surviving Hodgkin's disease following MOPP (nitrogen mustard, vincristine, procarbazine, and prednisone) therapy with and without radiotherapy. *Hum Genet*, **93**, 295-299.

Brinster, R. L. (2002) Germline stem cell transplantation and transgenesis. *Science*, **296**, 2174-2176.

Brinster, R. L. and Avarbock, M. R. (1994) Germline transmission of donor haplotype following spermatogonial transplantation. *Proceedings of the National Academy of Sciences of the United States of America*, **91**, 11303-11307.

Brinster, R. L. and Zimmermann, J. W. (1994) Spermatogenesis following male germ-cell transplantation. *Proceedings of the National Academy of Sciences of the United States of America*, **91**, 11298-11302.

Brinster, R. L. and Nagano, M. (1998) Spermatogonial stem cell transplantation, cryopreservation and culture. *Semin Cell Dev Biol*, **9**, 401-409.

Carson, S. A., Gentry, W. L., Smith, A. L. and Buster, J. E. (1991) Feasibility of semen collection and cryopreservation during chemotherapy. *Hum Reprod*, **6**, 992-994.

Chan, P. T. and Schlegel, P. N. (2000) Nonobstructive azoospermia. *Curr Opin Urol*, **10**, 617-624.

Clark, A. T., Bodnar, M. S., Fox, M., Rodriquez, R. T., Abeyta, M. J., Firpo, M. T. and Pera, R. A. (2004) Spontaneous differentiation of germ cells from human embryonic stem cells in vitro. *Hum Mol Genet*, **13**, 727-739.

Clouthier, D. E., Avarbock, M. R., Maika, S. D., Hammer, R. E. and Brinster, R. L. (1996) Rat spermatogenesis in mouse testis. *Nature*, **381**, 418-421.

Crabbe, E., Verheyen, G., Tournaye, H. and Van Steirteghem, A. (1997) The use of enzymatic procedures to recover testicular germ cells. *Hum Reprod*, **12**, 1682-1687.

Crabbe, E., Verheyen, G., Silber, S., Tournaye, H., Van de Velde, H., Goossens, A. and Van Steirteghem, A. (1998) Enzymatic digestion of testicular tissue may rescue the intracytoplasmic sperm injection cycle in some patients with non-obstructive azoospermia. *Hum Reprod*, **13**, 2791-2796.

Cremades, N., Bernabeu, R., Barros, A. and Sousa, M. (1999) In-vitro maturation of round spermatids using co-culture on Vero cells. *Hum Reprod*, **14**, 1287-1293.

Curaba, M., Verleysen, M., Amorim, C. A., Dolmans, M. M., Van Langendonckt, A., Hovatta, O., Wyns, C. and Donnez, J. (2010) Cryopreservation of prepubertal mouse testicular tissue by vitrification. *Fertil Steril*.

Daw, C., Neri, Q. V., Hu, J. C. Y., Schlegel, P. N., Rosenwaks, Z. and Palermo, G. D. (2011) Vitrification of testicular tissue: to mince. *Human Reproduction*, **26**.

Dobrinski, I., Avarbock, M. R. and Brinster, R. L. (1999) Transplantation of germ cells from rabbits and dogs into mouse testes. *Biology of reproduction*, **61**, 1331-1339.

Downs, K. M. and Davies, T. (1993) Staging of Gastrulating Mouse Embryos by Morphological Landmarks in the Dissecting Microscope. *Development*, **118**, 1255-1266.

Dym, M., He, Z., Jiang, J., Pant, D. and Kokkinaki, M. (2009) Spermatogonial stem cells: unlimited potential. *Reprod Fertil Dev*, **21**, 15-21.

Escalier, D., Gallo, J. M., Albert, M., Meduri, G., Bermudez, D., David, G. and Schrevel, J. (1991) Human Acrosome Biogenesis - Immunodetection of Proacrosin in Primary Spermatocytes and of Its Partitioning Pattern during Meiosis. *Development*, **113**, 779-&.

Evans, M. J. and Kaufman, M. H. (1981) Establishment in culture of pluripotential cells from mouse embryos. *Nature*, **292**, 154-156.

Foresta, C., Bettella, A., Marin, P., Galeazzi, C., Merico, M. and Scandellari, C. (2000) [Analysis of sperm aneuploidy in infertile subjects after chemotherapy treatment]. *Ann Ital Med Int*, **15**, 189-194.

Geens, M., De Block, G., Goossens, E., Frederickx, V., Van Steirteghem, A. and Tournaye, H. (2006) Spermatogonial survival after grafting human testicular tissue to immunodeficient mice. *Hum Reprod*, **21**, 390-396.

Geijsen, N., Horoschak, M., Kim, K., Gribnau, J., Eggan, K. and Daley, G. Q. (2004) Derivation of embryonic germ cells and male gametes from embryonic stem cells. *Nature*, **427**, 148-154.

Ginsburg, M., Snow, M. H. and McLaren, A. (1990) Primordial germ cells in the mouse embryo during gastrulation. *Development*, **110**, 521-528.

Guan, K., Nayernia, K., Maier, L. S., Wagner, S., Dressel, R., Lee, J. H., Nolte, J., Wolf, F., Li, M., Engel, W. et al. (2006) Pluripotency of spermatogonial stem cells from adult mouse testis. *Nature*, **440**, 1199-1203.

Hawkins, M. M. and Stevens, M. C. (1996) The long-term survivors. *Br Med Bull*, **52**, 898-923.

Heller, C. G. and Clermont, Y. (1964) Kinetics of the Germinal Epithelium in Man. *Recent Prog Horm Res*, **20**, 545-575.

Hochedlinger, K. and Plath, K. (2009) Epigenetic reprogramming and induced pluripotency. *Development*, **136**, 509-523.

Honaramooz, A., Snedaker, A., Boiani, M., Scholer, H., Dobrinski, I. and Schlatt, S. (2002) Sperm from neonatal mammalian testes grafted in mice. *Nature*, **418**, 778-781.

Hubner, K., Fuhrmann, G., Christenson, L. K., Kehler, J., Reinbold, R., De La Fuente, R., Wood, J., Strauss, J. F., 3rd, Boiani, M. and Scholer, H. R. (2003) Derivation of oocytes from mouse embryonic stem cells. *Science*, **300**, 1251-1256.

Hue, D., Staub, C., Perrard-Sapori, M. H., Weiss, M., Nicolle, J. C., Vigier, M. and Durand, P. (1998) Meiotic differentiation of germinal cells in three-week cultures of whole cell population from rat seminiferous tubules. *Biology of reproduction*, **59**, 379-387.

Hynes, R. O. (1992) Integrins: versatility, modulation, and signaling in cell adhesion. *Cell*, **69**, 11-25.

Izadyar, F., Pau, F., Marh, J., Slepko, N., Wang, T., Gonzalez, R., Ramos, T., Howerton, K., Sayre, C. and Silva, F. (2008) Generation of multipotent cell lines from a distinct population of male germ line stem cells. *Reproduction*, **135**, 771-784.

Jahnukainen, K., Ehmcke, J., Hergenrother, S. D. and Schlatt, S. (2007) Effect of cold storage and cryopreservation of immature non-human primate testicular tissue on spermatogonial stem cell potential in xenografts. *Hum Reprod*, **22**, 1060-1067.

Kanatsu-Shinohara, M., Ogonuki, N., Inoue, K., Miki, H., Ogura, A., Toyokuni, S. and Shinohara, T. (2003) Long-term proliferation in culture and germline transmission of mouse male germline stem cells. *Biology of reproduction*, **69**, 612-616.

Kerkis, A., Fonseca, S. A., Serafim, R. C., Lavagnolli, T. M., Abdelmassih, S., Abdelmassih, R. and Kerkis, I. (2007) In vitro differentiation of male mouse embryonic stem cells into both presumptive sperm cells and oocytes. *Cloning Stem Cells*, **9**, 535-548.

Keros, V., Hultenby, K., Borgstrom, B., Fridstrom, M., Jahnukainen, K. and Hovatta, O. (2007) Methods of cryopreservation of testicular tissue with viable spermatogonia in pre-pubertal boys undergoing gonadotoxic cancer treatment. *Hum Reprod*, **22**, 1384-1395.

Kubota, H., Avarbock, M. R. and Brinster, R. L. (2004) Culture conditions and single growth factors affect fate determination of mouse spermatogonial stem cells. *Biology of reproduction*, **71**, 722-731.

Kubota, H., Avarbock, M. R. and Brinster, R. L. (2004) Growth factors essential for self-renewal and expansion of mouse spermatogonial stem cells. *Proceedings of the National Academy of Sciences of the United States of America*, **101**, 16489-16494.

Kubota, H., Avarbock, M. R., Schmidt, J. A. and Brinster, R. L. (2009) Spermatogonial stem cells derived from infertile Wv/Wv mice self-renew in vitro and generate progeny following transplantation. *Biology of reproduction*, **81**, 293-301.

Kvist, K., Thorup, J., Byskov, A. G., Hoyer, P. E., Mollgard, K. and Yding Andersen, C. (2006) Cryopreservation of intact testicular tissue from boys with cryptorchidism. *Hum Reprod*, **21**, 484-491.

Lacham-Kaplan, O., Chy, H. and Trounson, A. (2006) Testicular cell conditioned medium supports differentiation of embryonic stem cells into ovarian structures containing oocytes. *Stem Cells*, **24**, 266-273.

Lawson, K. A. and Hage, W. J. (1994) Clonal analysis of the origin of primordial germ cells in the mouse. *Ciba Found Symp*, **182**, 68-84; discussion 84-91.

Loh, Y. H., Hartung, O., Li, H., Guo, C., Sahalie, J. M., Manos, P. D., Urbach, A., Heffner, G. C., Grskovic, M., Vigneault, F. et al.

(2010) Reprogramming of T cells from human peripheral blood. *Cell Stem Cell*, **7**, 15-19.

Martin, G. R. (1981) Isolation of a pluripotent cell line from early mouse embryos cultured in medium conditioned by teratocarcinoma stem cells. *Proceedings of the National Academy of Sciences of the United States of America*, **78**, 7634-7638.

Martin-du Pan, R. C. and Campana, A. (1993) Physiopathology of spermatogenic arrest. *Fertil Steril*, **60**, 937-946.

Mintz, B. (1957) Interaction between 2 Allelic Series Modifying Primordial Germ Cell Development in the Mouse Embryo. *Anat Rec*, **128**, 591-591.

Nagano, M., Avarbock, M. R. and Brinster, R. L. (1999) Pattern and kinetics of mouse donor spermatogonial stem cell colonization in recipient testes. *Biology of reproduction*, **60**, 1429-1436.

Nagano, M., McCarrey, J. R. and Brinster, R. L. (2001) Primate spermatogonial stem cells colonize mouse testes. *Biology of reproduction*, **64**, 1409-1416.

Nagano, M., Patrizio, P. and Brinster, R. L. (2002) Long-term survival of human spermatogonial stem cells in mouse testes. *Fertil Steril*, **78**, 1225-1233.

Nayernia, K., Nolte, J., Michelmann, H. W., Lee, J. H., Rathsack, K., Drusenheimer, N., Dev, A., Wulf, G., Ehrmann, I. E., Elliott, D. J. et al. (2006) In vitro-differentiated embryonic stem cells give rise to male gametes that can generate offspring mice. *Dev Cell*, **11**, 125-132.

Neri, Q. V., Fields, T., Rosenwaks, Z. and Palermo, G. D. (2012) A recreation of seminiferous tubule-like environment for spermatogonia stem cells. *Fertil Steril*, **98**, S143.

Neri, Q. V., Tanaka, N., Takeuchi, T., Rosenwaks, Z. and Palermo, G. D. (2006) Differentiation of embryonic stem cells into male germ cells. *Fertil Steril*, **86**, s107.

Neri, Q. V., Kollmann, Z., Chen, C., Rosenwaks, Z. and Palermo, G. D. (2009) A mouse model to restore spermatogenesis in pre-pubertal children undergoing chemotherapy. *Fertil Steril*, **92**, S66-S67.

Neri, Q. V., Tanaka, N., Takeuchi, T., Toschi, M., Rosenwaks, Z. and Palermo, G. D. (2006) Propagation and maturation of male gonocytes in vitro. *Fertil Steril*, **86**, s14.

Ogawa, T., Kita, K. and Kubota, Y. (2006) Proliferation of Spermatogonial Stem Cells and Spermatogenesis in vitro. *Reprod Med Biol*, **5**, 169-174.

Ogawa, T., Dobrinski, I., Avarbock, M. R. and Brinster, R. L. (1999) Xenogeneic spermatogenesis following transplantation of hamster germ cells to mouse testes. *Biology of reproduction*, **60**, 515-521.

Okada, H., Dobashi, M., Yamazaki, T., Hara, I., Fujisawa, M., Arakawa, S. and Kamidono, S. (2002) Conventional versus microdissection testicular sperm extraction for nonobstructive azoospermia. *The Journal of urology*, **168**, 1063-1067.

Orwig, K. E. and Schlatt, S. (2005) Cryopreservation and transplantation of spermatogonia and testicular tissue for preservation of male fertility. *J Natl Cancer Inst Monogr*, 51-56.

Orwig, K. E., Shinohara, T., Avarbock, M. R. and Brinster, R. L. (2002) Functional analysis of stem cells in the adult rat testis. *Biology of reproduction*, **66**, 944-949.

Palermo, G., Joris, H., Devroey, P. and Van Steirteghem, A. C. (1992) Pregnancies after intracytoplasmic injection of single spermatozoon into an oocyte. *Lancet*, **340**, 17-18.

Palermo, G. D., Neri, Q. V., Takeuchi, T. and Rosenwaks, Z. (2009) ICSI: where we have been and where we are going. *Semin Reprod Med*, **27**, 191-201.

Palermo, G. D., Schlegel, P. N., Colombero, L. T., Zaninovic, N., Moy, F. and Rosenwaks, Z. (1996) Aggressive sperm immobilization prior to intracytoplasmic sperm injection with immature spermatozoa improves fertilization and pregnancy rates. *Hum Reprod*, **11**, 1023-1029.

Parks, J. E., Lee, D. R., Huang, S. and Kaproth, M. T. (2003) Prospects for spermatogenesis in vitro. *Theriogenology*, **59**, 73-86.

Ramasamy, R., Yagan, N. and Schlegel, P. N. (2005) Structural and functional changes to the testis after conventional versus microdissection testicular sperm extraction. *Urology*, **65**, 1190-1194.

Ramasamy, R., Sterling, J., Fisher, E. S., Li, P. S., Jain, M., Robinson, B. D., Shevchuck, M., Huland, D., Xu, C. and Mukherjee, S. (2011) Identification of spermatogenesis with multiphoton microscopy: an evaluation in a rodent model. *The Journal of urology*, **186**, 2487-2492.

Rassoulzadegan, M., Paquis-Flucklinger, V., Bertino, B., Sage, J., Jasin, M., Miyagawa, K., van Heyningen, V., Besmer, P. and Cuzin, F. (1993) Transmeiotic differentiation of male germ cells in culture. *Cell*, **75**, 997-1006.

Reis, M. M., Tsai, M. C., Schlegel, P. N., Feliciano, M., Raffaelli, R., Rosenwaks, Z. and Palermo, G. D. (2000) Xenogeneic transplantation of human spermatogonia. *Zygote*, **8**, 97-105.

Robbins, W. A., Meistrich, M. L., Moore, D., Hagemeister, F. B., Weier, H. U., Cassel, M. J., Wilson, G., Eskenazi, B. and Wyrobek, A. J. (1997) Chemotherapy induces transient sex chromosomal and autosomal aneuploidy in human sperm. *Nat Genet*, **16**, 74-78.

Rohwedel, J., Guan, K. and Wobus, A. M. (1999) Induction of cellular differentiation by retinoic acid in vitro. *Cells Tissues Organs*, **165**, 190-202.

Russell, L. D. (1990) *Histological and histopathological evaluation of the testis*, Cache River Press.

Russell, L. D. and Brinster, R. L. (1996) Ultrastructural observations of spermatogenesis following transplantation of rat testis cells into mouse seminiferous tubules. *J Androl*, **17**, 615-627.

Saitou, M., Barton, S. C. and Surani, M. A. (2002) A molecular programme for the specification of germ cell fate in mice. *Nature*, **418**, 293-300.

Sasagawa, I., Yazawa, H., Suzuki, Y., Tateno, T., Ichiyanagi, O., Kobayashi, T., Matsuki, S. and Nakada, T. (2001) Reevaluation of testicular biopsies of males with nonobstructive azoospermia in assisted reproductive technology. *Arch Androl*, **46**, 79-83.

Schlatt, S., Kim, S. S. and Gosden, R. (2002) Spermatogenesis and steroidogenesis in mouse, hamster and monkey testicular tissue after cryopreservation and heterotopic grafting to castrated hosts. *Reproduction*, **124**, 339-346.

Schlatt, S., Honaramooz, A., Ehmcke, J., Goebell, P. J., Rubben, H., Dhir, R., Dobrinski, I. and Patrizio, P. (2006) Limited survival of adult human testicular tissue as ectopic xenograft. *Human Reproduction*, **21**, 384-389.

Schlegel, P. N. (2013) Perspectives on repeat sperm retrieval in nonobstructive azoospermia. *Fertil Steril*, **100**, 60.

Schmidt, J. A., Avarbock, M. R., Tobias, J. W. and Brinster, R. L. (2009) Identification of glial cell line-derived neurotrophic factor-regulated genes important for spermatogonial stem cell self-renewal in the rat. *Biology of reproduction*, **81**, 56-66.

Schover, L. R., Rybicki, L. A., Martin, B. A. and Bringelsen, K. A. (1999) Having children after cancer. A pilot survey of survivors' attitudes and experiences. *Cancer*, **86**, 697-709.

Schrader, M., Muller, M., Straub, B. and Miller, K. (2002) Testicular sperm extraction in azoospermic patients with gonadal germ cell tumors prior to chemotherapy--a new therapy option. *Asian journal of andrology*, **4**, 9-15.

Schrader, M., Heicappell, R., Muller, M., Straub, B. and Miller, K. (2001) Impact of chemotherapy on male fertility. *Onkologie*, **24**, 326-330.

Seki, T., Yuasa, S., Oda, M., Egashira, T., Yae, K., Kusumoto, D., Nakata, H., Tohyama, S., Hashimoto, H., Kodaira, M. et al. (2010) Generation of induced pluripotent stem cells from human terminally differentiated circulating T cells. *Cell Stem Cell*, **7**, 11-14.

Shinohara, T., Avarbock, M. R. and Brinster, R. L. (1999) beta1- and alpha6-integrin are surface markers on mouse spermatogonial stem cells. *Proceedings of the National Academy of Sciences of the United States of America*, **96**, 5504-5509.

Shinohara, T., Inoue, K., Ogonuki, N., Kanatsu-Shinohara, M., Miki, H., Nakata, K., Kurome, M., Nagashima, H., Toyokuni, S., Kogishi, K. et al. (2002) Birth of offspring following transplantation of cryopreserved immature testicular pieces and in-vitro microinsemination. *Hum Reprod*, **17**, 3039-3045.

Siu, M. K. Y. and Cheng, C. Y. (2004) Dynamic cross-talk between cells and the extracellular matrix in the testis. *Bioessays*, **26**, 978-992.

Siu, M. K. Y. and Cheng, C. Y. (2004) Extracellular matrix: Recent advances on its role in junction dynamics in the seminiferous epithelium during spermatogenesis. *Biology of reproduction*, **71**, 375-391.

Sofikitis, N., Ono, K., Yamamoto, Y., Papadopoulos, H. and Miyagawa, I. (1999) Influence of the male reproductive tract on the reproductive potential of round spermatids abnormally released from the seminiferous epithelium. *Hum Reprod*, **14**, 1998-2006.

Sousa, M., Cremades, N., Alves, C., Silva, J. and Barros, A. (2002) Developmental potential of human spermatogenic cells co-cultured with Sertoli cells. *Hum Reprod*, **17**, 161-172.

Staerk, J., Dawlaty, M. M., Gao, Q., Maetzel, D., Hanna, J., Sommer, C. A., Mostoslavsky, G. and Jaenisch, R. (2010) Reprogramming of human peripheral blood cells to induced pluripotent stem cells. *Cell Stem Cell*, **7**, 20-24.

Staub, C. (2001) A century of research on mammalian male germ cell meiotic differentiation in vitro. *J Androl*, **22**, 911-926.

Stefankiewicz, J., Kurzawa, R. and Drozdzik, M. (2006) [Environmental factors disturbing fertility of men]. *Ginekol Pol*, **77**, 163-169.

Steinberger, A., Steinberger, E. and Perloff, W. H. (1964) Mammalian Testes in Organ Culture. *Exp Cell Res*, **36**, 19-27.

Steinberger, E., Steinberger, A. and Perloff, W. H. (1964) Studies on Growth in Organ Culture of Testicular Tissue from Rats of Various Ages. *Anat Rec*, **148**, 581-589.

Surveillance, Epidemiology, and End Results (SEER) Program (www.seer.cancer.gov) Research Data (1973-2012), National Cancer Institute, DCCPS, Surveillance Research Program, Surveillance Systems Branch, released April 2015, based on the November 2014 submission.

Tadokoro, Y., Yomogida, K., Ohta, H., Tohda, A. and Nishimune, Y. (2002) Homeostatic regulation of germinal stem cell proliferation by the GDNF/FSH pathway. *Mechanisms of development*, **113**, 29-39.

Takahashi, K. and Yamanaka, S. (2006) Induction of pluripotent stem cells from mouse embryonic and adult fibroblast cultures by defined factors. *Cell*, **126**, 663-676.

Takeuchi, T., Neri, Q. V. and Palermo, G. D. (2008) Male gamete empowerment. *Ann N Y Acad Sci*, **1127**, 64-66.

Tanaka, A., Nagayoshi, M., Awata, S., Mawatari, Y., Tanaka, I. and Kusunoki, H. (2003) Completion of meiosis in human primary spermatocytes through in vitro coculture with Vero cells. *Fertil Steril*, **79 Suppl 1**, 795-801.

Tanaka, H. and Baba, T. (2005) Gene expression in spermiogenesis. *Cell Mol Life Sci*, **62**, 344-354.

Tanaka, N., Feliciano, M., Rosenwaks, Z. and Palermo, G. D. (2005) Isolation of spermatogonial stem cells. *Fertil Steril*, **84**, S75-S75.

Tanaka, N., Feliciano, M., Rosenwaks, Z. and Palermo, G. D. (2005) A simple method for selection of spermatogonial stem cells. *Hum Reprod*, **20**, i46.

Tanaka, N., Neri, Q. V., Takeuchi, T., Rosenwaks, Z. and Palermo, G. D. (2006) Selecting and propagating spermatogonial stem cells in a serum-free medium. *Hum Reprod*, **21**, i208.

Tanaka, N., Takeuchi, T., Schlegel, P. N., Feliciano, M., Rosenwaks, Z. and Palermo, G. D. (2004) A specific membrane marker identifying spermatogenic cells. *Fertil Steril*, **82**, S184-S185.

Tesarik, J. (2004) Overcoming maturation arrest by in vitro spermatogenesis: search for the optimal culture system. *Fertil Steril*, **81**, 1417-1419.

Tesarik, J., Bahceci, M., Ozcan, C., Greco, E. and Mendoza, C. (1999) Restoration of fertility by in-vitro spermatogenesis. *Lancet*, **353**, 555-556.

Tesarik, J., Nagy, P., Abdelmassih, R., Greco, E. and Mendoza, C. (2002) Pharmacological concentrations of follicle-stimulating hormone and testosterone improve the efficacy of in vitro germ cell differentiation in men with maturation arrest. *Fertil Steril*, **77**, 245-251.

Thomson, J. A., Itskovitz-Eldor, J., Shapiro, S. S., Waknitz, M. A., Swiergiel, J. J., Marshall, V. S. and Jones, J. M. (1998) Embryonic stem cell lines derived from human blastocysts. *Science*, **282**, 1145-1147.

Tournaye, H., Goossens, E., Verheyen, G., Frederickx, V., De Block, G., Devroey, P. and Van Steirteghem, A. (2004) Preserving the reproductive potential of men and boys with cancer: current concepts and future prospects. *Hum Reprod Update*, **10**, 525-532.

Toyooka, Y., Tsunekawa, N., Takahashi, Y., Matsui, Y., Satoh, M. and Noce, T. (2000) Expression and intracellular localization of mouse Vasa-homologue protein during germ cell development. *Mechanisms of development*, **93**, 139-149.

Wright, V. C., Chang, J., Jeng, G. and Macaluso, M. (2006) Assisted reproductive technology surveillance--United States, 2003. *MMWR Surveill Summ*, **55**, 1-22.

Wright, V. C., Schieve, L. A., Reynolds, M. A., Jeng, G. and Kissin, D. (2004) Assisted reproductive technology surveillance--United States, 2001. *MMWR Surveill Summ*, **53**, 1-20.

Wyns, C., Van Langendonckt, A., Wese, F. X., Donnez, J. and Curaba, M. (2008) Long-term spermatogonial survival in cryopreserved and xenografted immature human testicular tissue. *Hum Reprod*, **23**, 2402-2414.

Wyns, C., Curaba, M., Martinez-Madrid, B., Van Langendonckt, A., Francois-Xavier, W. and Donnez, J. (2007) Spermatogonial survival after cryopreservation and short-term orthotopic immature human cryptorchid testicular tissue grafting to immunodeficient mice. *Hum Reprod*, **22**, 1603-1611.

Ying, Y. and Zhao, G. Q. (2001) Cooperation of endoderm-derived BMP2 and extraembryonic ectoderm-derived BMP4 in primordial germ cell generation in the mouse. *Developmental biology*, **232**, 484-492.

Ying, Y., Qi, X. and Zhao, G. Q. (2001) Induction of primordial germ cells from murine epiblasts by synergistic action of BMP4 and BMP8B signaling pathways. *Proceedings of the National Academy of Sciences of the United States of America*, **98**, 7858-7862.

Yoshimizu, T., Obinata, M. and Matsui, Y. (2001) Stage-specific tissue and cell interactions play key roles in mouse germ cell specification. *Development*, **128**, 481-490.

Zapzalka, D. M., Redmon, J. B. and Pryor, J. L. (1999) A survey of oncologists regarding sperm cryopreservation and assisted reproductive techniques for male cancer patients. *Cancer*, **86**, 1812-1817.

Zhao, G. Q., Deng, K., Labosky, P. A., Liaw, L. and Hogan, B. L. (1996) The gene encoding bone morphogenetic protcin 8B is required for the initiation and maintenance of spermatogenesis in the mouse. *Genes & development*, **10**, 1657-1669.

Zhao, G. Q., Chen, Y. X., Liu, X. M., Xu, Z. and Qi, X. (2001) Mutation in Bmp7 exacerbates the phenotype of Bmp8a mutants in spermatogenesis and epididymis. *Developmental biology*, **240**, 212-222.

Made in United States
Orlando, FL
04 November 2026

90200978R00071